The Edible Crab
and its Fishery in British Waters

The Edible Crab

and its Fishery in British Waters

Eric Edwards Ph.D., M.I.Biol.

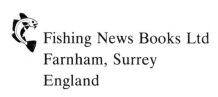

Fishing News Books Ltd
Farnham, Surrey
England

This is a Buckland Foundation Book, one of a series providing permanent record of annual lectures maintained by a bequest of the late Frank Buckland.

First Published 1979

British Library CIP Data

Edwards, Eric
   The edible crab and its fishery in
   British waters. – (Buckland Foundation.
   Books).
   1. Crab fisheries – Great Britain
   I. Title II. Series
   338.3'72'53842   SH380.45.G7

   ISBN 0–85238–100–X

Text set in 11/12 pt Photon Times, printed by photolithography, and bound in Great Britain at The Pitman Press, Bath

Contents

List of figures

List of plates

List of tables

11

Foreword

The Buckland Foundation exists to keep alive the memory of the Founder, Frank Buckland, and, with money he left, to promote a series of lectures on subjects in the field of fisheries on which he worked with such notable effect. Crabs qualify without difficulty because Buckland was concerned about their conservation and the need for some order and regulation in their fishery. Although fishermen in his time were no doubt as familiar with the grounds they worked as their counterparts are today, there was no background of reliable scientific information about growth, breeding, distribution, migration and behaviour of crabs. We have come to recognize that such knowledge provides the only sound basis for regulation of the fishery and effective conservation of the stocks. This basic need is now well understood and appreciated by forward-looking fishermen.

The application of modern technology has made inshore fishing vessels more efficient and has increased their power and range. More pots can be carried by crab fishermen and this may increase fishing pressure on local stocks. On the shore, processing facilities are being steadily improved and a variety of attractive products is being produced. Export possibilities are being developed. Nevertheless, Dr Edwards shows that in some areas there are large resources of crabs not fully exploited, although in others additional protective measures are needed if yields are to be maintained or increased. The biology of crabs is not uniform throughout their range and regional differences in regulations governing their fishery will be needed if the best results are to be obtained.

All these questions, and many others, are examined by the author, the facts being presented in straightforward terms backed wherever possible by his own observations or by an analysis of the results of other scientific work. Much new scientific material is presented. The result is a volume of lasting value, written in a way that anyone concerned with the shellfish industry, whether as a fisherman, processor, merchant, scientist or administrator, can readily understand. It shows where the industry stands, what it has done, where expansion is possible, what problems remain to be solved and where further scientific research is needed.

13

At a time such as this, when the minds of all concerned with sea fisheries are anxiously reviewing the probable outcome of the long struggle to work out a sensible Common Fisheries Policy within the European Economic Community, the book is very timely. But it is more than an examination of current problems: it presents a record of sustained scientific investigation over many years of the biology of the edible crab. There is much new information and a reassessment of earlier work. It is a substantial biological monograph in its own right. It fulfills admirably the intentions of the Foundation.

H. A. COLE
Chairman, Trustees of the Buckland Foundation

Plate 1   The European edible crab *Cancer pagurus*

1 Introduction

The edible crab (*Cancer pagurus*) is the only species of crab regularly fished for human consumption in European waters. This crab is particularly abundant in the coastal waters of northwest Europe, being present in large numbers in areas where the sea bed is rocky. Although showing a preference for this rugged type of bottom, *Cancer pagurus* is also found on other bottoms ranging from sand to pure mud.

Although edible crabs are landed in most European countries, the most important crab fisheries are found off Norway, around the coasts of Scotland and along the northeast and southwest coasts of England. A major fishery for *Cancer pagurus* also occurs around France, mainly off Brittany. Smaller fisheries are found off Spain, Portugal and around Ireland, where stocks have only recently been exploited (*Table 1*).

Table 1. *Quantities of crabs (metric tons) landed in various European countries 1969–1973*

Country	1969	1970	1971	1972	1973
Belgium	134	228	246	292	325
Denmark	—	—	—	—	—
France	8,900*	9,750*	11,843*	11,362*	—
Republic of Germany	19	11	13	20	29
Ireland	605	642	904	946	953
Netherlands	18	33	46	66	68
Norway	2,707	2,670	2,291	2,432	3,029
Spain	14	14	—	16	—
Sweden	—	67	57	51	67
England and Wales	4,190	3,728	4,139	4,522	4,772
N Ireland	2	1	2	1	1
Scotland	2,389	2,250	1,706	1,897	2,250

Figures from: *Bulletin Statistique des Pêches Maritimes.*

* Source: Assemblée Général du Comité Central des Pêche Maritimes – data include spider crabs (*Maia*)

In Britain, one of the main crab fishing countries in Europe, about 6,500 tonnes of edible crabs with a first-sale value of about £1,500,000 are landed every year. This figure represented 45 per cent of the total value of British shellfish landings during each of the past five years (1972–76). The main British crab fishing grounds are along the east coast of Scotland and England, where about 25 per cent of the total annual catch is landed. Other important fishing areas are along the Devon coast, around Cornwall and off Orkney and Shetland (*Fig 1*).

Fig 1 Principal crab fishing ports in Britain and Ireland

In Britain, crabs are caught in baited traps, the design of which varies around the coast. Most of the catch is sold to processing factories, sited at certain ports, where the crabs are cooked, the meat extracted and used in the preparation of crab paste or sold (either canned or frozen) for hotel and home consumption.

Bearing in mind the economic importance to Britain of the crab fisheries, it is not surprising that most of the biological studies on the edible crab *Cancer pagurus* have been carried out by British scientists. This included the work of Professor A Meek, who studied crabs off the Northumberland coast from 1895 to 1925, and Dr H C Williamson employed by the Scottish Fishery Board at Aberdeen during the same period. Both these scientists added considerably to our knowledge of the biology and behaviour of the crab.

More recently extensive studies on the British crab stocks have been undertaken by staff at the Fisheries Laboratory, Burnham-on-Crouch, Essex, and the Marine Laboratory, Aberdeen. This work included studies on the growth and migrations of crabs along the northeast coast of Britain and in the English Channel. This publication describes the biology and behaviour of the European edible crab, aspects of its fishery and the measures used to manage the stocks. Many of the comments and conclusions reached have resulted from the cooperation of fishermen and merchants who catch and handle crabs. Although their names are too many to list individually their support and assistance is gratefully acknowledged.

2 History of the fishery and its regulation

Early fishery

The European edible crab has been exploited for many centuries. The Roman 'carabus' probably refers to *Cancer pagurus* and suggests that this species was of domestic importance to them. There are, however, no exact records of the early fishery for crabs, although passing references to crab boats are to be found as early as the 12th century in the Rolls of the Whitby Abbey, Yorkshire.

An early record of a crab fishery in Norfolk is mentioned in 'A Guide about Cromer' published in 1800 by Edmond Burtell who states: 'Lobsters, crabs, whiting, cod-fish and herring are all caught here (Cromer) in the finest perfection'. It is also recorded that at that time Cromer had a considerable trade in corn and crabs with London. In 1874 Holdsworth published a comprehensive account of the various fisheries around the British Isles and includes some reports on crab fishing. According to Holdsworth 'Crab and lobster fishing employ a large number of men at particular seasons. They are caught in a creel or cage trap, also called a 'pot'.'

In 1875 Frank Buckland, the eminent marine biologist of Victorian times, made a detailed study of the crab, lobster and herring fisheries of Norfolk. His report contains useful information about the limits of the crab fishery, the number of boats employed at the two major ports, Sheringham and Cromer, and the state of the fishery. Particular emphasis was given to the fact that there had been a considerable decrease in crab landings in the previous year (1874). The cause was attributed to the wholesale destruction of small crabs and the killing of both crabs and lobsters carrying spawn. According to Buckland small crabs had been used for bait or sold for food at a price of between $1\frac{1}{2}$d and $2\frac{1}{2}$d for 20 since the early 1800s. Buckland also calculated that about 750,000 small crabs were brought ashore and sold every month at Norfolk ports; this affected the stocks because it is reported that after 1868 the supply of crabs from Norfolk ports to Billingsgate, London, virtually ceased. Many Norfolk fishermen were concerned about this destruction, and at Cromer they even imposed their own voluntary restriction to prohibit the destruction of small crabs less than $4\frac{1}{4}$ inches (108 mm) across the carapace.

As a result of his enquiry Frank Buckland came to the conclusion that these fisheries along the Norfolk coast should be protected by statutory restrictions, and following his recommendations to Parliament the *1876 Crab and Lobster Fisheries (Norfolk) Act* was introduced. This imposed a restriction on the possession or offer for sale of crabs measuring less than $4\frac{1}{4}$ inches or lobsters of less than 7 inches. The Act also restricted the sale of crabs or lobsters carrying spawn. Norfolk fishermen were therefore instrumental in bringing about the introduction of the first statutory regulations for the management of a crab fishery. Surprisingly, these regulations were applied only to this one fishery in the country.

Buckland continued his studies on the crab fisheries of the country, and a comprehensive report prepared by Buckland and his colleague Walpole was published in 1877 describing the crab and lobster fisheries of England, Wales, Scotland and Ireland. This report provides a considerable amount of information on the English crab fisheries of the time, including extensive details on the fishery and life history of the crab following visits by the two biologists to about 27 ports in various parts of the country in 1876 and 1877. At each port a public enquiry was held and witnesses presented evidence which was recorded; this included the price of crabs in various areas, the fluctuations in fishermen's catches, local information on the crab's biology and also methods of fishing at that time.

The 1877 report states that the principal crab fisheries in England were in the counties of Northumberland, Yorkshire, Norfolk, Sussex, Hampshire, Devon and Cornwall. In Scotland large numbers of crabs were landed at ports along the northeast coast. Meetings were held at Montrose, Aberdeen, Peterhead and Fraserburgh and from the account written it is obvious that crab fishing was important in this area.

The chief market for fish and shellfish at that time was Billingsgate Market, London, the inland towns being then supplied direct from there. In earlier years all the crabs were taken to London by welled smacks or steamers but the construction of railways enabled the 'great towns' to be supplied direct from the coast. From all accounts the crab fishery at this time was an important one, although details of sea fisheries landings were not recorded until 1886 and so no records of crab landings are available prior to that date.

This enquiry also accumulated considerable knowledge on the natural history of crabs. Buckland pointed out that there were two different sizes of edible crabs around the British Isles; the largest crabs were found on the coasts of Devon and Cornwall, while the crabs along the northeast coast of Scotland and England were small. This difference in size was attributed by Buckland to the influence of the Gulf Stream on the southwest coast of England.

Legislation

Following their study of the British crab and lobster fisheries, Buckland and Walpole made certain recommendations to Parliament and on September 1, 1877 the *Fisheries (Oysters, Crabs and Lobsters) Act 1877* was passed which introduced Statutory Regulations for the whole of England, Wales and Scotland, thus replacing the 1876 Norfolk Act. The new regulations included a minimum landing size of $4\frac{1}{4}$ inches (108 mm) across the broadest part of the carapace, the protection of crabs carrying spawn and the protection of soft-shelled crabs, *ie* those which had recently cast their shells. The minimum size for lobsters was set at 8 inches (203 mm) total length and the Board of Trade was invested with powers to enforce, after a public enquiry, local close seasons for crabs and lobsters in any area.

In 1894, the *Sea Fisheries Regulation Act* authorized local Sea Fisheries Committees to impose further by-laws for the protection of the crab fishery. These Committees introduced local bye-laws regarding the taking of crabs and therefore there was some variation between the eleven Sea Fisheries areas. For example, along the east coast of England the tendency has been to protect soft-shelled and under-sized crabs, while in parts of the south and west, where crabs are generally larger, the legal landing size was raised to 5 inches (127 mm).

The *1877 Fisheries (Oysters, Crabs and Lobsters) Act* enforcing a national $4\frac{1}{4}$ inch size limit remained in operation until 1951 when agreement was reached on the need for some further protective legislation. This was introduced with the *Sea Fishing Industry (Crabs and Lobsters) Order 1951,* which increased the national minimum size of crabs to $4\frac{1}{2}$ inches (115 mm) shell width. In areas where a local Sea Fisheries Committee by-law of 5 inches was enforced no change was made.

In 1966 the *Sea Fishing Industry (Crabs and Lobsters) Order 1951* was repealed by the *Sea Fishing Industry (Crabs and Lobsters) Order 1966*. This new Order introduced certain changes regarding the national lobster legislation but retained on a national basis the $4\frac{1}{2}$ inch legal size limit for the edible crab, *Cancer pagurus*. In

March 1976 under the *Immature Crabs and Lobsters Order* this size was converted to an equivalent shell width of 115 mm.

The *Sea Fish (Conservation) Act 1967* also repealed the *1877 Fisheries (Oysters, Crabs and Lobsters) Act.*

Statutory restrictions affecting the present day crab fishery are shown below:

Under Section 17 of the *Sea Fisheries (Shellfish) Act 1967* it is an offence for any person to take, have in his possession, sell, expose for sale, buy for sale, or consign to any person for the purpose of sale:

(*a*) any edible crab carrying spawn attached to the tail or other exterior part of the crab, or

(*b*) any edible crab which has recently cast its shell.

Under the *Immature Crabs and Lobsters Order 1976* it is an offence to land, sell, expose or offer for sale or be in possession of for the purpose of sale any crabs of the species *Cancer pagurus* which are smaller than 115 mm (4.5 inches) measured across the broadest part of the back.

Further restrictions relating to crabs have been imposed by various Sea Fisheries Committees, but they relate mainly to the use of undersized crabs for bait (Eastern Sea Fisheries Committee) or to a closed season from November 1 to December 31 (Northumberland Sea Fisheries Committee) aimed at protecting soft-shelled crabs.

Fishing and gear

As far as it is known baited pots for catching crabs have been in regular use since the 1860s. Prior to that time the main method of collection was by either hand picking or netting.

Bell (1853) describes how lobsters and crabs were often taken in considerable numbers from clefts in the rock at low water mark, using a gaff. Holdsworth (1874) reported that pots made of 'withy' (willow) were used in Cornwall (*Fig 2*) but describes in detail the hoop nets, also known as 'trunks', which were widely used in those days to catch both lobsters and crabs. These hoop (*Fig 2*) nets were constructed of an iron ring about two foot across, to which was attached a shallow net bag of about two inch mesh; they were baited with fish, sunk to the bottom and hauled at regular intervals. Their main disadvantage,

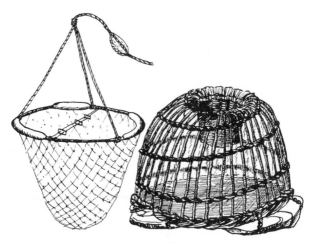

Fig 2 Hoop net and a traditional willow ink-well pot used in the
early 1800s

however, was that they needed constant attention and there was a
limit to the number which could be managed.

Towards the second half of the nineteenth century the use of pots
became more widespread. Buckland and Walpole (1877) provide the
most comprehensive information on the fishing gear used during the
period: '... The fishery for crabs and lobsters is conducted in every
part of the Kingdom in the same manner. Traps made in wicker-work
or of a wicker frame covered in netting and usually known as 'pots' or
'creels' are almost universally used. The pot is baited with some fish,
fresh for crabs and stinking fish for lobsters and sunk in three fathoms
to 45 fathoms water ... '. Buckland's description of the crab pot used
in 1875 along the east coast of England shows that it was similar to
that in use today, except that it had a slightly larger base. According
to Buckland and Walpole (1877), pots had then only recently been introduced in some areas; prior to this, hoop nets were widely used.
Buckland also reported that this method of fishing needed constant
attention and each boat only worked between 30 and 40 hoops.
Buckland (1875) in his report on the Norfolk crab and lobster
fisheries, mentioned that crab pots had been introduced into that area
about 12 years earlier, in 1862 or 1863. There is no record of where
they came from but an old Sheringham fisherman told the author that
his grandfather told him that the first crab pot used in Norfolk came
from the Yorkshire coast. After the introduction of pots the number of

hoop nets used declined rapidly, but up to the time of the First World War they were occasionally used for lobster fishing early in the spring in both Yorkshire and Norfolk. Today they are not used for crab fishing in these areas but are sometimes still employed to take lobsters off the Suffolk and Essex coasts, *eg* on the West Rocks and the Roughs off Walton-on-the-Naze and Harwich.

Boats

Since the early part of the last century there has also been very little change in the design of boats used to fish for crabs around Britain.

A description of the boats used in the past by Sheringham and Cromer fishermen is given by Holdsworth (1874). The boats described were similar in design to those in present use, being double-ended and broad-beamed but propelled by oars and sail. At that time there were about 100 boats at Sheringham and 50 at Cromer, each crewed by two men, who worked about 25 to 35 pots per boat. In the present day fishery, the boats (*Plate 2*) are now motorized and some 30 boats work a total of 4,000 pots in the Cromer and Sheringham area.

Buckland and Walpole (1877) refer to the coble which is still widely used for crab and lobster fishing along the northeast coast of England between Holy Island in Northumberland and Flamborough Head in

Plate 2 Norfolk beach boats at Sheringham

23

Yorkshire. This open boat (*Plate 3*) is believed to have originated from the Viking long-ships which raided these coasts, and it is ideally designed for launching from exposed beaches in an area where beaches and cliff clefts often serve as havens. The coble is between 20 and 30 ft long and has a high bow and a deep forefront, which facilitates easy launching off steep beaches. Originally designed as an open sail and oar boat, the coble (*Fig 3*) needed only a slight modification before being motorized, and following the Great War (1914–18) petrol engines were installed. No sailing cobles now remain, and at the present time the majority are fitted with diesel engines of 20–40 hp.

Although cobles have been used for centuries in the east coast crab and lobster fisheries, larger vessels have also been used for lining and drifting, as well as potting. In 1833, the first yawl was built at Scarborough and this type of vessel gradually replaced the 'five man boats' which had been used for decades for trawling along the Yorkshire coast. During the last quarter of the nineteenth century a vessel known as the 'mule' was used at Whitby and Scarborough; this hybrid craft (hence the name) combined features of both the coble and the yawl and was between 30 and 50 ft long. Crewed by four or five men, it was first used for herring drifting but it was eventually used for crab and

Plate 3   Cobles at Whitby, Yorkshire

24

Fig 3 A Yorkshire sailing coble, from a model in the science museum, London

lobster fishing and for lining. By all accounts the mule's sea-going ability was not exceptional, and being an open boat conditions for the crew were not easy. During the early 1920s the Scottish-designed 'fifie' boat widely used for inshore fishing at Scottish northeast coast ports was introduced, and this gradually replaced the mules along the northeast coast of England. Built in Scotland for Yorkshire owners the fifies were known locally as 'keel boats' and were fully decked with a wheel-house situated aft. Ranging in length from 40 to 50 ft and fully motorized, these vessels were ideal for the various fishing activities practised along this coast. The present day keel boats have a similar design but have lengths up to 55 ft and are fitted with depth sounders and other navigational aids. Many of these boats today fish for crabs and lobsters around England and are crewed by four or five men who can handle between 400 and 500 pots daily.

The inshore fishing boats worked in the south and west coast of England in the 19th century varied considerably in design but most had a square transom stern, thus providing a marked contrast with the double-ended boats so common along the east coast. Around the 1800s various types of luggers, such as the Mount's Bay type, fished for crabs under sail. These two-masted wooden boats were rigged with a large dipping lug-sail on the foremast and a standing lugsail on the mizzen, above which a topsail was hoisted. From this type of vessel has developed the typical Cornish crab lugger, which ranges in size

25

from 30 to 50 ft. These vessels are now all powered, and fitted with echo-sounders and other electronic equipment.

Landings

By referring to earlier figures it is possible to plot the growth of the crab fishery. The collection of commercial statistics of the sea fisheries of the UK was first undertaken in 1886 and few accurate data on landings are available before this date.

In 1888, following the introduction of the *Sea Fisheries Regulation Act,* local Sea Fisheries Committees were encouraged by the Board of Trade to record the quantities of both fish and shellfish landed at ports in their area. Records of crab landings are available for most of the country from 1850 onwards. These can be extracted from the Minutes or Reports of the various Sea Fisheries Committees or from the Government *Sea Fisheries Statistical Tables.* Scottish landings are published in the *Annual Reports* of the *Fishery Board of Scotland* and in more recent years in the *Scottish Sea Fisheries Statistical Tables.* Both the English and Scottish statistical tables are published annually by HM Stationery Office.

Crab landings from 1890 to 1953 were all recorded as numbers caught, but from 1954 onwards all figures given in the statistical tables are shown as weights landed – in units of hundredweights (112 pounds or 51 kg).

The annual crab landings in England and Wales during the period 1904–1953 varied between approximately 2,780,000 and 8,120,000 crabs. During this time the landings and value gradually increased (*Table 2*) but it is impossible to determine whether this was the result of an increase in fishing effort or an increase in the crab stocks. It can be seen that the average landing value for crabs increased steadily except during the period 1924–1933. The low figure for this period was due to the economic recession of the late twenties when the price of crabs fell to a very low level. No marked improvement occurred until about 1939.

During the period 1895–1955 the main crab fishing areas were found along the east coast of England, particularly in Norfolk, Northumberland and Yorkshire. It is only in comparatively recent years that the southwest crab fisheries have increased in importance. Up to 1969 the major crab fishery was in the area along the Yorkshire coast, between the Tees and the Humber. Gradually the southwest

26

area increased in importance, and in the 1970s over half the country's crab catch was landed in ports in Devon.

Table 2 Crab landings in England and Wales during the years 1904–1953 (numbers) and 1954–1969 (weights)

Average for 10-year period	Average landing	Average value (£)
	(Nos)	
1904–13	4,924,433	56,642
1914–23	4,489,012	67,727
1924–33	5,931,175	67,526
1934–43	5,044,660	81,164
1944–53	6,234,540	243,831
	(cwt)	
1954–63	67,518	271,238
1964–69	58,564	312,351

3 The present day fishery

Landings and value

In recent years crabs have become one of the most important sources of income to UK shellfishermen. During the last 10 years the average annual catch has been around 5,000 tonnes, with a first-sale value of £700,000 (*Table 3* and *Fig 4*).

Table 3 Crab landings and their value in Scotland, England and Wales 1967–1976

Year	Weight (cwt)		Total UK (cwt)	Value
	England and Wales	Scotland		
1976	112,000	40,000	152,000	£1,917,036
1975	97,000	33,000	130,000	£1,332,000
1974	71,000	46,000	117,000	£1,199,000
1973	94,000	45,000	139,000	£1,041,000
1972	89,000	37,000	126,000	£758,000
1971	81,000	34,000	115,000	£670,000
1970	73,000	44,000	118,000	£586,000
1969	82,000	47,000	129,000	£595,000
1968	54,000	45,000	99,000	£455,000
1967	55,000	34,000	89,000	£409,000

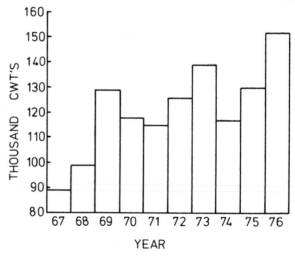

Fig 4   Annual recorded crab landings for the UK 1967–1976

Ports in England land the highest crab catches; an average 75 per cent of the UK crab catch is landed in England, landings at Welsh ports are minimal (*Table 4*) and the remainder (20 per cent) is landed in Scotland.

Table 5 shows the proportion of the crab catch landed in different areas of England and Wales. It can be seen how important the east coast of England has been for crabs over the 55 years (1904–59) when about 80 per cent of the country's catch was landed at ports in Northumberland, Yorkshire and Norfolk.

Since 1960 an increasing proportion of the country's crab catch has been landed in the south coast area, which includes the important Devon fishery. By the 1970s at least half the country's catch came from this area.

Table 4 Top ten crab ports 1975

	England and Wales	Scotland
1	R Dart (Kingswear)	Fraserburgh
2	Salcombe	Gairloch
3	Plymouth	St Abbs
4	Weymouth	Hoy
5	Selsey	Burnmouth
6	Bridlington	Westray
7	Cromer	Stromness
8	Sheringham	Dunbar
9	Filey	Rousay
10	Littlehampton	Gourdon

Table 5 Proportion of the crab catch landed in three different fishing areas of England and Wales, 1904–1975

Period	East Coast (*Berwick–Deal*)	South Coast (*Dover–Isle of Scilly*)	West Coast (*Sennen Cove–Silloth*)
1904–13	81%	16%	3%
1914–23	81%	16%	3%
1924–33	85%	12%	3%
1934–43	89%	9%	2%
1944–53	90%	9%	1%
1954–59	78%	21%	1%
1960–69	66%	33%	1%
1970–75	31%	68%	1%

In Scotland there are important crab fisheries around Shetland and Orkney (landing 30 per cent of the catch) while the east coast has important centres at Fraserburgh, Anstruther, Leith and Eyemouth. The east coast ports of Scotland land about 60 per cent of the Scottish crab catch and only about 10 per cent of the country's catch comes from the west coast. Stocks there are, however, reported to be abundant and not fully exploited.

Season

In Britain some boats fish for crabs throughout the year, but the main season is from April to November. On the east coast of England peak catches occur in May and June but in the southwest of England catches are low in the early part of the year and high from September to November (*Fig 5*). In Scotland the most important months for crabs are May, June, July and August but some landings are made thoughout the year.

Crab landings in all areas are affected by various factors, such as weather or crab behaviour – including moulting and migrations – and by the seasonal emphasis on the capture of other fish species. In most areas lobsters are caught with the same type of traps, and since lobsters command a much higher price than crabs, there is at certain times a seasonal emphasis on lobsters. This occurs particularly in the summer when calm weather allows the pots to be set very close inshore on rich lobster ground. In addition, during the months of July and August crabs are moulting and a large proportion of the catch is soft shelled and has to be rejected, and so fishermen concentrate on lobsters and avoid the crab grounds. Also, some boats, particularly the large type, will only fish for crabs when catches are good during the spring and autumn and will turn to alternative fishing such as trawling, lining or trammel-netting during the remaining months of the year. Economic factors can also affect the overall landings from a crab fishery; for example a reduction in the crab landings from the Yorkshire fishery during the 1960s was due to a greater emphasis on trawling from certain Yorkshire ports, rather than to a decline in the crab stocks.

The fishing gear

Crabs are captured in baited traps, whose design varies around the

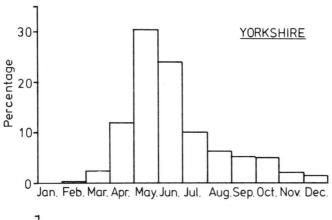

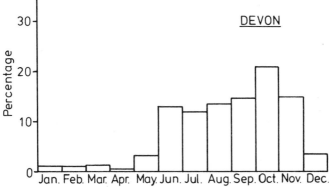

Fig 5 Monthly distribution of the crab catches landed in Yorkshire and Devon during 1975

coast. These 'pots', as they are most usually called, are designed to allow entry of the crab but to hinder its escape. In an industry that is distributed along the whole of the British coastline, tradition and individual preferences have influenced the pattern of trap design and therefore a wide variety of pot types is found.

The two main types of traps used are the creel (*Fig 6*) fished mainly on the east coast of England and all round Scotland and the ink-well pot which is mainly used in southwest England and in Wales. Creels are constructed of wood and netting and usually have a base diameter of 24 × 18 inches and a height of 12 inches. Three half hoops of hazel or cane support the netting of either sisal or twine or the now widely used synthetic fibres, such as nylon or 'Courlene'. The pot has two

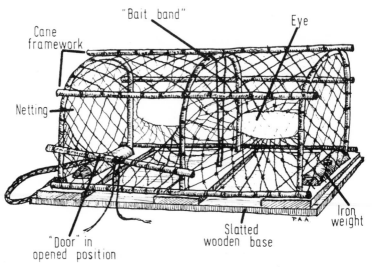

Fig 6 A Yorkshire crab and lobster creel

openings ('eyes'), each constructed in the form of a short funnel and entering the pot from opposite sides, each eye having a diameter of about 5 inches at the narrowest part. Pieces of iron lashed to the wooden base are used to weight the pot. The bait is held between toggles in the single 'bait band', which is a double length of stretched twine fitted from the roof to the wooden base. These pots are usually made by the fishermen themselves. Variation in creel design is common; for example the Norfolk creel (*Fig 7*) has a cast iron base (known locally as a 'music') and two bait bands, and the catch is removed through the roof of the pot which opens. The Norfolk creels also have four hoops rather than the three used in the Scottish and north of England creels (Edwards 1965a).

The Scottish creel is similar to that used on the northeast coast of England but a number of variations are at present in use around the Scottish coasts. On the east coast of Scotland the creel is generally somewhat larger than that described and on the west coast the creels often have only one eye. The creel is also used in Orkney and Shetland. In these areas stones are often lashed inside the creel to act as a sinker. Thomas (1958) has described in detail the Scottish crab fisheries and the gear used.

Circular ink-well pots have been used in southwest England and in part of Wales for centuries. In days gone by these were made from withy cut mainly from specially planted willow garths, but, today,

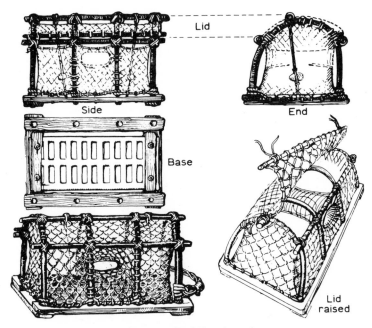

Fig 7 A Norfolk crab creel

although the shape of the pot has changed little, durable modern materials are used (*Plate 4*). In the 1960s these pots were made of a wire framework, bound by chestnut hoops. This framework was covered with netting; the entrance or neck was made of basket-work and the pot of weldmesh. The bait was held in place by wooden 'skivers' pushed through the bait and into the basket-work neck. Pots of this type were made by the fishermen for their own use and few were offered for sale.

Today (1978) the situation has changed. Although ink-well pots of wire and netting are still in use, they are being steadily replaced by welded tubular polythene frame pots or 'plastic pots' as they are called. This development has mainly occurred in the Devon area, but it is gradually spreading along the English Channel, including the Cornish coast. The 'plastic pots' were originally designed and produced by Mr Nantes of Weymouth. They are usually sold as frames and the fishermen cover them with netting. Basket work necks have now largely been replaced by ready-made plastic or glass-fibre necks. With the change has come the use of a rubber band (usually old car inner tubes) placed around the bottom of the neck to hold the bait firmly in place.

33

Plate 4   Removing the catch from a Devon ink-well pot

This change to more modern pots has led to a subsidiary industry springing up making pots, pot frames, pot entrances and swivels to prevent the gear getting tangled. All this makes the crab fishery in southwest England more important as a source of employment than is generally realised.

In the 1960s many Cornish fishermen adopted the barrel-type pot (*Plate 5*) used by French fishermen who worked off our southwest coasts, mainly for crawfish. No doubt, the first barrel pots used were French ones washed ashore during storms and repaired. This pot constructed in the form of a cylinder of wooden lathes, with the two ends enclosed by netting, is said by experienced fishermen to be economical and easy to construct, and to make catches which compare with those of the traditional pots. However, barrel pots were easily damaged by storms, especially when fishing in shallow waters. The effectiveness of

Plate 5   Barrel pots at Hayle, Cornwall

the plastic ink-well pot has meant that very few barrel pots are now in use in southwest England; those that are left are mainly used around Cornwall.

Most crab fishermen make and repair their own traps but manufactured ones are becoming more popular and are becoming available in various styles. Most of these are based on traditional designs but often incorporating modern materials such as plastic and fibre glass in their construction. It seems likely that as the seasonal occupations of crab fishermen become more diverse and their spare-time ashore becomes more reduced, the purchase of ready-made traps will assume greater importance than at the present time.

Bait

The catching power of a trap depends primarily on the attraction of the bait. Almost any kind of fish can be used for bait, but price and availability often influence the choice. Gurnards which have a very tough skin and last well in the trap are very popular, whereas herring, although attractive to crabs, are not widely used since they soon become soft and break up. In some areas fish offal such as cod heads, plaice 'frames' and the backbones of skate are often used because of their relative cheapness. Fresh dogfish are also considered a good bait

35

for crabs. Crab fishermen stress that to catch crabs fresh bait is required – the use of salted bait results in low crab catches.

Whenever possible the bait should be changed every day; even though what was in use previously may appear untouched much of the attractive juices may be lost and the bait will not be fully effective. Some fishermen believe that when working one ground continuously it is advantageous to change, at intervals, to different types of bait. Although certain commercial companies in the USA have produced and marketed artificial baits none are available on the British market. The development of a cheap chemical attractant which could be used to catch crabs (and lobsters) would be a promising advance. Tests at the Institute of Marine Biochemistry, Aberdeen using chemical substances were reasonably effective but the materials were too expensive for commercial use and the project has been dropped. However, the use of fish extract, which is cheaper to produce, could have potential value.

Fishing methods

The traps are worked in strings, or fleets (*Plate 6*), of between 20 and 70 pots set on the sea bottom at intervals along a main rope which has a buoyed anchor or sinker attached to each end. The number of traps worked in each fleet will depend on the size of the boat and on the local conditions. A fleet should not have more creels than can comfortably be accommodated and baited *etc* on the boat. For example, a 30 foot boat may work about 200 traps, whereas a 50 foot boat can handle up to 500. The pots are usually secured to a 'back rope' of about 1½ inches (38 mm) circumference, at intervals of about 8–10 fathoms (15–18 metres). The spacing is important as it is preferable that not more than two traps are off the bottom together during hauling.

When fishing in strong currents, on exposed coasts or in deep water, and particularly when large fleets of pots are used, it is usual to moor the whole fleet to anchor. Sometimes the 'end' pots are more heavily weighted. In some areas the weight of the traps alone is sufficient to hold them on the bottom.

Dan buoys painted in a distinctive colour or with a flag may be used to mark the end of the fleets. The buoy line may be lighter than the back ropes and generally has a length three times the depth of water. Small boats sometimes fish with individually-buoyed traps, but

36

Plate 6   Ready to shoot a string of creels on a Whitby Keel boat (note dan buoy)

this method is more usual when fishing for lobsters on carefully selected ground.

When shooting the pots they should be laid on a straight course. As far as possible the back rope should be kept tight to help spread the pots over the ground and to lessen the amount of movement. The pots are usually shot with the vessel steaming with the tide, although some fishermen prefer to shoot at an angle to the tide.

The traps are usually set on selected ground and left fishing overnight and hauled the next day if weather allows. Sometimes the pots are hauled for a second time within 24 hours but this practice is more common in lobster fishing. After hauling the catch is removed, the traps rebaited and the gear re-set again (*Plate 7*). At one time the traps were hauled by hand, today however, mechanical capstans (*Plate 8*), worked off the main engine, are widely used to haul the strings of traps. This has helped to increase the number of grounds that can be

Plate 7 Emptying and rebaiting Yorkshire crab pots

worked because areas with strong tides and deep water can now be exploited.

In recent years most fishermen have changed over from capstans to hydraulic line haulers. These lift the traps clear of the water and above the level of the vessel's side, so that they only need to be swung inboard. This takes much of the backache out of 'lifting in' and makes it possible to work effectively with one less crewman.

It is of interest to record that in the Norfolk crab fishery all pots were hauled by hand until as late as the 1960s. During the 1967 season one of the Cromer boats fitted a hydraulic capstan and over the next few years there was a gradual change from hand-hauling to hydraulic hauling. By 1970 nearly all the fleet of 40 crab boats in this fishery were fitted with capstans (Brown 1975).

Crab fishing occurs at varying depths and distance from the shore. In Scotland and in northeast England crabbing is usually within 12 miles of the coast, generally between two and six miles. This only involves daily excursions to the fishing grounds which may be found just outside the harbour entrance or up to 10 miles from port.

In other parts of the country crab fishing can take place a long way offshore; for example, the Devon crab fishery was originally concentrated close to the shore. Most fishermen shot their pots within six to eight miles of the coast. Gradually, during the last 10 years, there has

Plate 8   Mechanical haulers are used to lift the pots

been a tendency to set pots further from the shore. This is an attempt to make larger, better quality catches in order to meet the costs of bigger and better equipped boats. At present most of the crabs taken in this fishery are caught 25–35 miles off Start Point. There are, however, problems about working on the offshore grounds partly due to the exposed nature of the area and the risk of damage to the pots by trawlers of several nationalities.

The majority of men employed in the crab fishery are full-time professional fishermen, although they may not spend the whole year crabbing. Some fishermen, particularly the older ones, work their traps only during the spring and summer and lay their boats up for the remainder of the year. Some of the fishermen will continue to work within the fishing industry during the winter on other types of fishing, making or repairing pots, or baiting long-lines.

In some areas people with full-time employment ashore are purchasing small boats and going crab fishing in their spare time. These hobby fishermen usually only work a few traps but their numbers are increasing as working hours of their main employment are reduced.

The boats used are usually skipper-owned and are sometimes crewed by members of the same family. Large boats work on a share basis, with perhaps several of the crew owning part of the gear. In addition, several of the larger shellfish merchants now own and operate their own boats.

Boats of a variety of classes and of overall lengths between 12 and 55 ft are used to catch crabs. Apart from individual preference, the main factors determining the size of boat are the availability of suitable harbours and the existence of alternative types of fishing in places, where for certain reasons, such as exposure, traps cannot be worked at all seasons. Half-decked motor boats of from 30 to 35 ft length are popular, although in the Devon fishery, where offshore grounds are worked, vessels of 45–50 ft are common. The Cornwall Sea Fisheries Committee bans shellfishing by boats of 50 ft and over in their District.

Regulations

As described earlier in this account, national and local regulations govern the size and condition of crabs which can be landed and marketed. This means that all crabs must be carefully examined and selected after being removed from the traps; this is done while at sea. In all areas of Britain under the *Immature Crab and Lobster Order 1976* crabs smaller than 115 mm (4.5 inches) must be returned to the sea. However, along the coasts of Hampshire, Dorset and Cornwall local Sea Fisheries Committee Regulations have raised the legal minimum size to 5 inches (127 mm). In addition, national regulations prohibit the landing of 'berried' (*ie* egg-bearing) females and soft-shelled crabs which have moulted and are in poor condition. These regulations are conscientiously enforced in most parts of the country by local Fishery Officers, and fishermen who contravene them may be fined. There is, however, no national closed-season for crabs or restriction on the number or type of traps fished.

Handling crabs at sea

Live crabs should be handled as little as possible after capture. They should not be pulled from the traps by their claws, since these are

40

readily shed, resulting in a multilated specimen. Experienced crab fishermen do not have the crabs lying on deck exposed to sun and wind, where they rapidly become weakened and die, but stow them away properly as soon as they are caught either into boxes, barrels or into the bottom of the boat. Crabs should be packed back uppermost and put close together to reduce movement and to prevent fighting. While at sea, wet sacks are often placed over the boxes to keep the crabs moist. During hot weather buckets of clean seawater are often poured over the crabs to keep them alive.

Commercial utilization

After being removed from the traps the day's catch is packed alive into boxes or baskets and the crabs are sold either by auction at the port of landing or by a pre-arranged agreement with a merchant or processor. The merchants then cook the crabs by boiling and distribute them to inland wholesalers. In some areas the crabs are sent in large quantities to processing factories where the meat is extracted for freezing and canning. In contrast some fishermen boil their own catches and have stalls on the seafront of seaside towns to sell crabs and other shellfish to holidaymakers.

In Britain the catch is usually sold by weight in units of pounds or stones (14 pounds) and prices fluctuate according to the supply and demand. In 1976 the average price ranged from 9p to 11p per pound (all in) on the northeast coast of England and 10p to 13p per pound for hen crabs in Devon; cock crabs fetched 25p per pound. It is, however, of interest to record that in the Norfolk fishery crabs at ports like Cromer and Sheringham are sold by number rather than by weight, in units of two, known as a 'cast', so that a daily catch of 180 crabs would be 90 casts. A large unit – 'one long hundred' is 240 crabs or 120 casts. This method of counting crabs has been used for decades in Norfolk and was mentioned by Buckland in his 1875 report.

Originally, all crabs were sold freshly cooked in the shell but the market for boiled crabs was limited by the lack of refrigeration facilities and the rather poor keeping quality of crab meat. During recent years improved transportation and freezing have allowed the market to expand, but the demand for whole crab is still rather limited due mainly to the difficulty experienced by the general public in extracting the meat from the shell. However, the development of processing factories, where the crabs are boiled in large quantities and the meat extracted and packeted, has expanded the sales of crab meat and thus stabilized the fishery. For example in southwest England it has been estimated that 90 per cent of all crabs landed pass through processing plants.

At these plants, pickers remove the meat from the shell by hand; at the time of writing mechanical methods of extracting crab meat have not been used in Britain although evaluations are being made of US equipment which may be of use to the British industry. While most of the crabs are broken up and the meat extracted, for the Scandinavian market crabs are normally washed, cooked and frozen whole, female crabs being required for this particular trade. Some crabs landed in the Shetlands are flown to Sweden and sold alive.

Live crabs do not store or travel well and it is necessary, therefore, to have processing facilities within reasonable distance of the ports of landing. In Britain most of the major crab processing plants are on the coast, *eg* at Paignton, Newlyn, Plymouth, Hull, Berwick, Stromness, Lerwick, Inverberrie and Gairloch.

Processing crabs

Various steps are necessary to process the edible crab. The description given here is the technique recommended by the Torry Research Station, Aberdeen (see *Catching, Handling and Processing Crabs,* Torry Advisory Note No 26 Edwards & Early 1967).

Killing crabs
Crabs should be alive at the start of processing but must be killed prior to boiling, otherwise they will shed their claws and legs. They can be killed either by 'drowning' in fresh water or by spiking the nerve centre or 'brains'.

Crabs immersed in fresh water at about 50°F (10°C) will die in 3 to 5 hours; in fresh water at about 100–120°F (38–49°C) the time may be as little as 30 minutes. This method is much more practical than spiking if large quantities of crab are being handled. The resulting meat also seems to have a better texture than if the crabs are spiked. Spiking is often used to kill large cock crabs and this is done by inserting a pointed rod just above the mouth parts; the crab is killed almost immediately. This method is recommended by The Universities Federation for Animal Welfare as being more humane than drowning.

Boiling crabs

Crabs should be cooked immediately after killing by boiling them in water containing 2–3 per cent salt for 20–30 minutes, depending on size. Small northeast coast and Scottish crabs usually only need 20 minutes boiling while the larger southwest England crabs need 25–30 minutes. This amount of cooking will kill a very high proportion of the bacteria present, and will destroy any potential disease-causing organisms.

When boiling crabs the times given above will be effective only if the water is at boiling point 212°F (100°C) throughout. Bubbling of the water surface in a boiler does not necessarily mean that all the water is at boiling point; boiling can begin at the top of the boiler before the water at the bottom has reached that temperature. This problem often occurs in incorrectly regulated steam-jacketed boilers. The Torry Research Station always recommends that boilers should be fitted with a thermometer; the sensitive bulb should be in that part of the boiler that takes longest to heat. Each boiler should also have a timer, so that errors in boiling times can be avoided, particularly where a number of boilers are in use together.

Cooling cooked crabs

On removal from the boiler, the crabs, preferably contained in wire-mesh keeping baskets, should be first hosed down with cold fresh water and then left to cool at room temperature. The heaps of crabs should not be too large or cooling will be slow. This part of the cooling process is known as 'keeping', and three to four hours are necessary to allow the meat to coagulate or set. If the meat is removed too soon after boiling there is a loss of yield. Torry recommends that if the keeping time has to be longer than four hours, for instance storage overnight, then the crabs should be kept in a chill room.

Crabs which are to be sold whole should now be selected – only specimens with two claws and all the eight legs being used. The remainder of the crabs are then taken to the picking area for removal of the meats.

Picking the meats

The muscle or white meat is removed principally from the claws, and sometimes also from the body and walking legs (*Fig 8*); the brown meat, which consists mainly of digestive gland ('liver') and reproductive organs is removed from the main body, sometimes called

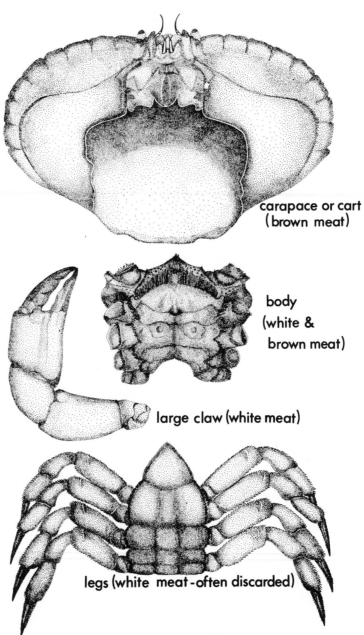

carapace or cart
(brown meat)

body
(white &
brown meat)

large claw (white meat)

legs (white meat - often discarded)

Fig 8 Parts of the edible crab

44

carapace or 'cart'. Hand picking is most usual, although compressed air jets are sometimes used to assist in extracting the meat from the body core, or from the walking legs.

The technique of picking does vary slightly according to the operator but the following description will provide a guide:

First the body is pulled away from the 'cart' (upper shell) and the brown meat scraped out with a spoon. The claws are then broken into segments, and each segment is cracked with a heavy rod; the white meat is then picked out with a short knife or specially designed tool. Body meat is removed from the leg sockets after the legs have been removed using the handle of a short spoon. Sometimes the walking legs of the larger crabs are cracked open and the white muscle meat extracted but the legs of small crabs are usually too small to handle and are dumped.

When crabs are being handled in bulk, it is better to have operators extracting meat from all parts of the crabs simultaneously. However, it is not uncommon for a day's catch to be boiled first and then the claws are removed and frozen, only the bodies being dealt with that day.

An effective method is to have one or more operators removing the claws from the crabs and then other persons either removing the brown meat from the bodies or breaking open the claws.

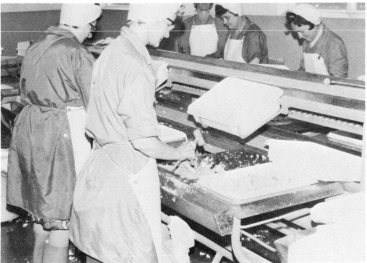

Plate 9 Picking crab meat by hand *Crown copyright* (*DAFS*)

After extraction, the brown meat, which tends to vary in colour from crab to crab, is minced finely to give it a uniform appearance and then packed ready for freezing. The white meat from the claws can either be extracted in large chunks or in smaller pieces. Both types of meat should be packed and frozen without delay. Much of the crab meat produced in this country is used in the preparation of crab paste which has a ready market. In recent years, however, more crab meat has been sold in attractive packs frozen ready for hotel or home consumption.

The amount of meat extracted from crabs can vary tremendously with the season and the fishing ground. On average, good quality crabs can yield about 30 per cent of the total body weight and about two thirds of this will be brown body meat. A higher quantity of white meat is extracted from male crabs which have larger claws, while more brown meat can be extracted from the body of a female crab. Female crabs yield less total meat than a male of a similar shell size.

In commercial practice yields ranging from 20–25 per cent are usual. The presence of soft-shelled crabs in the landed catches can reduce the overall yield and processors should exclude these before boiling.

Tests have shown that the total meat yield from the edible crab varies throughout the season in relation to the moulting and breeding

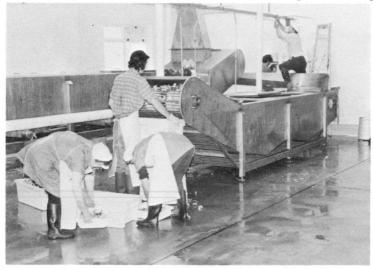

Plate 10   A modern crab boiling plant *Crown copyright* (*DAFS*)

cycle. This variation particularly applies to female crabs due to changes in the brown meat content following development of the gonads which in a fully-ripe female fill the whole carapace with a red granular tissue known in the trade as 'coral'. The yield of white meat from the claws of both males and females shows less seasonal variation, in fact there is a fairly constant meat yield representing about 30 per cent of the claw weight.

The Devon crab fishery is mainly based on hen crabs taken during the months of September to December when they are full of brown body meat prior to spawning (*see* Reproduction Chapter 4). These crabs have a high meat yield and are particularly suitable for the Swedish market.

On the northeast coast of England, on average crabs gave their highest meat yield in July but females had a high brown meat yield content in October and November, again due to the development of gonads or 'coral' prior to spawning.

4 Reproduction

The reproductive cycle in the edible crab has been studied in detail by several British scientists, including Dr Williamson and Professor Meek, who both made extensive investigations from 1895 to 1918. Most of the observations made were on crab stocks found along the east coast of Britain, particularly in Scotland and Northumberland. Dr Marie Lebour, working at the Marine Laboratory at Plymouth between the two World Wars, added considerably to the knowledge of the larval stages of the crab and collected information on the general life-history of this crustacean. Pearson (1908) also described the reproductive structures in crabs (*Fig 9*).

More recently the author has made certain observations on the reproductive cycle in *Cancer pagurus* which add to the available knowledge. Much of this work was done on the east coast of England (Norfolk and Yorkshire) and around the coast of southern Ireland (Edwards 1966b, Edwards & Meaney 1968).

Reproductive structures
The sex of a crab can easily be determined by the external features – the female having a broader abdomen than the male (*Fig 10*). In addition, in a mature male the claws are larger than those of a mature female of a similar size. The dorsal side of the male's carapace is also much flatter than the female's which has a 'humped' appearance. In the male, abdominal appendages are present only on the first and second abdominal somites and these are modified to form copulatory organs. The female differs by having all the somites freely movable and there is one pair of appendages on each of the second, third, fourth and fifth somites – these form the 'swimmerets' to which the eggs are attached. The external genital openings of the female consist of a pair of large openings situated on the sternum of the sixth thoracic somite.

Maturity
The male
The problem of recognizing maturity in male crustacea has been the subject of much research in recent years. In general the most effective method

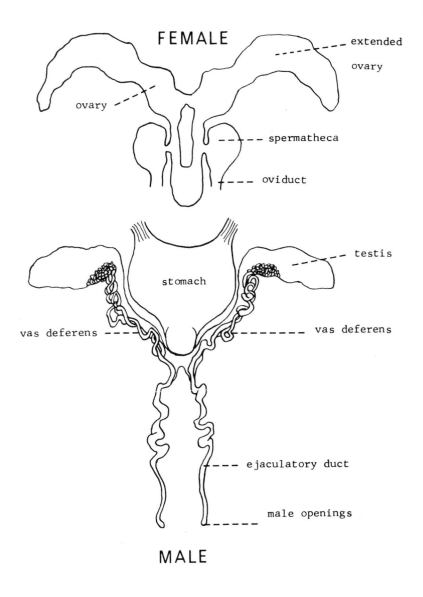

Fig 9 The male and female reproductive structures in *Cancer pagurus* (after Pearson, 1908)

49

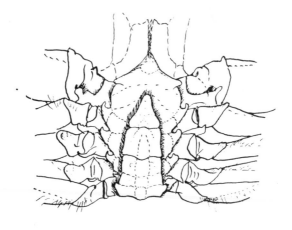

MALE

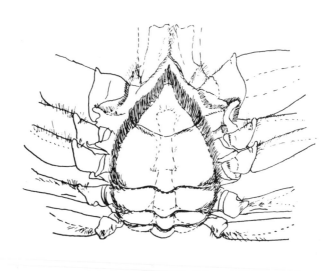

FEMALE

Fig 10   External abdominal features of male and female *Cancer pagurus*

is to dissect the crab and to examine the condition of the testis. In the immature male crab the testis is extremely small and is transparent in colour. In a mature specimen the testes are swollen and they may cover almost the whole of the digestive gland. The vasa deferentia – a pair of long convoluted tubes passing from each testis to the sex organs – also become swollen in the mature crab. This swelling is due to the presence of masses of spermatophores (sperm sacs) which also gives rise to the white appearance of the testis and vas deferens so characteristic of the ripe male.

Laboratory observations on male crabs caught in Yorkshire waters showed that in the immature male the vas deferens is small and difficult to locate due to its pale colour. In this condition the vasa deferentia contain few spermatophores (*Plate 11*) and are filled with a colourless fluid in which are numerous minute corpuscles of fat. As recorded by Williamson (1900) the vas deferens in the mature male is easily identified by its white swollen appearance. In addition to this feature, a more advanced maturity stage is reached when the sperm

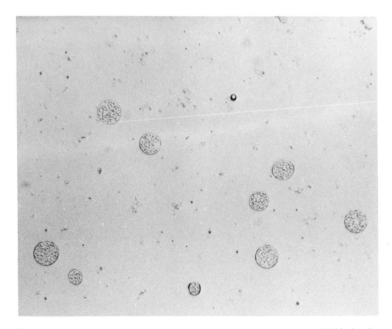

Plate 11   Smear from a testis of an immature male (90 mm carapace width) showing very few spermatophores × 1,000

duct becomes unusually swollen. Microscopic examination of the contents of the vasa deferentia from ripe male crabs showed numerous spermatophores enclosing the developing spermatozoa. The minute fat corpuscles found in the immature specimens were also present (*Plate 12*).

Size at maturity. Based on the presence or absence of ripe vasa deferentia, observations by the author and earlier workers suggest that most male crabs over 110 mm carapace width (4.3 inches) are sexually mature.

Table 6 shows the results of an examination of 160 males caught off Yorkshire and 182 off the southwest coast of Ireland. The results from both areas confirm that most crabs over 110 mm shell width are mature, but some crabs mature at a smaller size.

Observations showed that male crabs were mature during all seasons of the year although the main period of moulting and copulation is during the summer months. Soft-shelled crabs (*ie* those which have recently undergone a moult) were also found to have mature

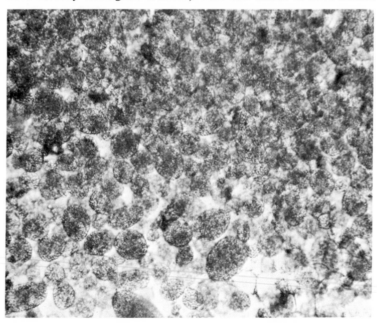

Plate 12   Smear from a testis of a mature male (126 mm carapace width) showing mass of spermatophores × 1,000

Table 6 Proportion of hard-shelled mature male crabs in various size groups

Size group mm	Total number examined	Number with white swollen vasa deferentia	% mature
Yorkshire			
80–89	15	1	6.6
90–99	29	5	17.2
100–109	50	32	64.0
110–119	43	40	93.0
120>	23	23	100.0
Totals	160	101	63.1
SW Ireland			
80–89	25	0	Nil
90–99	32	5	15.6
100–109	43	30	69.8
110–119	54	50	92.6
120>	28	28	100.0
Totals	182	113	62.1

sperm present in the spermatophores and it appears that the moulting process does not affect the stage of maturity.

Secondary sexual characteristics. In addition to using direct reproductive development as a guide to maturity, there is evidence to indicate that the development of the claws as a secondary sexual characteristic can give a guide to maturity in various species of crabs. Studies by the author showed that an enlargement of the claws occurred in male *Cancer pagurus* at a size of about 110 mm carapace width when there was an increase in the relative growth of the large claws. This is a development of one of the male crab's secondary sexual characteristics.

These observations support the conclusion that the critical puberty moult has been achieved when males reach a carapace width of 110 mm; at this size they would be approximately three years old.

Little is known about the development of sperm in male crabs. Pearson (1908) described how the spermatozoa produced in the testis collect in the vas deferens in groups and become surrounded by a capsule to form a spermatophore (*Plate 13*). As stated earlier the vas deferens of mature males is full of spermatophores enclosing millions of sperm. Little work has been done on the structure of the crab sperm but in 1965 the author examined sperm under an electron microscope.

53

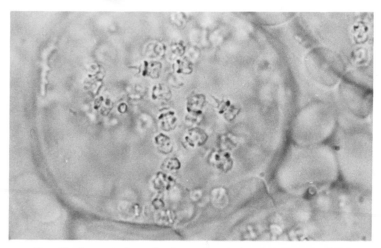

Plate 13 Spermatophore of a ripe male enclosing spermatozoa × 2,000

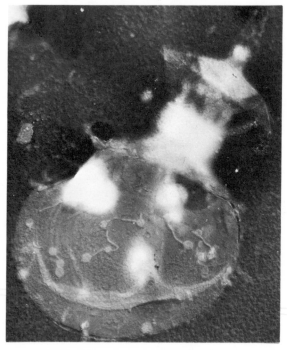

Plate 14 Electron micrograph of a mature spermatozoa taken from the oviduct of an impregnated female × 19,600

54

The study showed that mature sperm taken from impregnated females had a 'head' and 'tail' typical of the 'true' sperm (*Plate 14*).

The female

In females, in contrast to males, maturity cannot be accurately determined by the state of the gonads. This is because development of the gonads sometimes occurs only a considerable time after mating and hence at copulation the ovaries can still be immature.

For this reason maturity in the female crab is best identified by three main characteristics:

    (*a*) the presence of eggs on the abdomen
    (*b*) the presence of sperm in the spermatheca
    (*c*) the ripeness of the ovaries

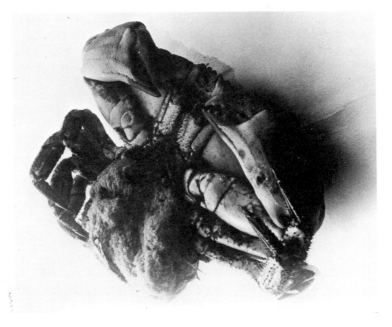

Plate 15   A berried female

Berried females. Various scientific workers have used the presence of eggs on the female crustacea to determine maturity. This has been done with lobsters, spider crabs and king crabs off Alaska all of which are recorded in the egg-carrying stage in traps in the commercial fisheries. It is of interest to record that around Britain egg-bearing female crabs, or 'berried' crabs (*Plate 15*) as they are usually called, are rarely taken in the commercial traps but are sometimes caught by trawlers working offshore grounds. During the author's studies along the Yorkshire coast (1961–1966) a total of 23,000 females of over 115 mm (4.5 inches) shell width were examined and only 200 of these were found to be carrying eggs. The smallest berried female in this sample had a shell width of 129 mm (5.1 inches) and the majority were over 152 mm (6 inches) width (*Table 7*). These observations suggest that in Yorkshire egg-carrying in *Cancer pagurus* commences at a size of about 130 mm (5 inches) shell width, although it is possible that some smaller crabs do also carry eggs. Dr Williamson (1900) working in Scotland reported a small berried crab of 115 mm (4.5 inches) but stated that out of 65 berried crabs examined, 37 were over 152 mm (6 inches). From these rather limited data he concluded that in Scotland most of the crabs between 127 mm and 152 mm (5 and 6 inches) shell width were mature. Pearson (1908), who examined crabs at Port Erin, believed that many crabs did not bear their first batch of eggs until reaching 152 mm (6 inches). The available information does however suggest that most females are mature when they reach a size of 127 mm (5 inches) shell width.

Table 7 Percentage size distribution of 200 berried female crabs taken in traps off the Yorkshire coast, 1961–1966

Carapace width

Inches	4½/5	5/5½	5½/6	6/6½	6½/7	7/7½	7½/8	8/8½	8½/9
mm	115	127	140	152	165	178	191	203	216
	0	3%	7%	17%	20%	31%	12%	9%	1%

Throughout the literature conflicting reports have appeared in an attempt to account for the absence of berried crabs in the traps in the commercial fishery. As early as 1898 Cunningham reported that very few berried crabs were ever seen by fishermen and he concluded that the large egg mass on the abdomen restricted their ability to enter the

56

traps. Williamson (1900) also referred to the\
berried crabs seen during his studies in Scottish wa\
he considered that the egg mass could restrict ent\
pointed out that there was evidence to show th\
sometimes moved considerable distances offshore to sp\
the experiments he carried out in tanks at Dunbar, Scot\
that while carrying eggs female crabs remained half-b\
sand, ate little and their movements were limited. M ⟨1904)
collected records of the crab catch at Seahouses, Northumberland,
and reported that, during a period of six years, although 54,178
marketable-sized crabs were examined only 133 were berried. He concluded ' . . . that the great distention of the abdomen by the burden of
the eggs renders it very difficult for the female to enter the crab pot.'

Other workers have also suggested that egg-bearing females of
various species of crab are more timid than other members of the population. Observations by the author on berried crabs kept in the tanks
at the Fisheries Laboratory, Burnham-on-Crouch showed that none
ever entered a baited trap placed in a tank, although males and non-
berried females readily entered the trap and ate the bait. This observation suggests that berried females avoid this type of fishing gear. This
could be due to timidity or a protective behaviour towards the eggs, or
simply, inability to climb through the entrance of the trap.

Hatching females. Even though berried females are rarely taken in the
commercial fishery, females which have recently hatched their eggs are
taken in considerable numbers (*Table 7*). Crabs which have been
carrying eggs can be recognised by the presence of empty egg capsules
left on the swimmerets of the abdomen. These crabs also have a
characteristic appearance, their abdomen being dirty and discoloured,
and their shells are usually heavily encrusted with epifauna such as bar-
nacles, saddle oysters (*Anomia*) and tube worms. The discoloured
appearance of the body is believed to be due to their habit of partially
burying themselves in the sea bed when carrying eggs. The encrusted condition of the carapace is probably due to a reduction in the number of
moults – a condition often occurring in egg-bearing female crabs.

During the Yorkshire and Irish studies the author examined large
numbers of female crabs for the presence of egg capsules. The observations made suggest that the main mass of empty egg capsules
was lost shortly after the eggs hatched but in most cases a few capsules were left adhering to the swimmerets, indicating that the crab

le 8 Monthly occurrence of egg bearing and hatching female crabs, Yorkshire (1963) and southwest Ireland (1969 and 1970)

Yorkshire

Month	Total females examined	Numbers carrying eggs	Numbers with egg remains	% hatching
March	—	No	Fishing	—
April	520	0	13	2.5
May	763	1	54	7.1
June	700	2	84	12.0
July	1,060	4	144	13.6
August	729	1	189	25.9
September	520	0	58	11.2
October	300	0	5	1.7
November	250	0	0	—
December	—	No	Sample	—
Totals	4,842	8	547	11.3%

Southwest Ireland

Month	Total females examined	Numbers carrying eggs	Numbers with egg remains	% hatching
March	—	No	Fishing	—
April	—	No	Fishing	—
May	402	2	4	1.0
June	495	1	156	31.5
July	412	3	145	35.2
August	414	0	97	23.4
September	331	0	25	7.6
October	488	0	19	3.9
November	—	No	Sample	—
December	—	No	Sample	—
Totals	2,542	6	446	17.5%

Note: Includes only crabs of 115 mm carapace width and over.

58

had recently carried eggs.

The collected data on the sizes of these hatching crabs provided additional evidence regarding size at maturity. In both southwest Ireland and Yorkshire most of the hatching females had shell widths between 152 mm and 190 mm (6–7 inches). None were recorded below 5 inches shell width.

Presence of sperm. Williamson (1904), studying the female sex organs in *Cancer pagurus,* described how a fluid, produced from the glands lining the spermatheca, flowed into the vagina when the male copulatory organ was introduced. On withdrawal of the male organ this fluid hardened on contact with seawater to form a white structure which plugged the opening of each oviduct (*Plate 16*). The same author reported that this 'sperm plug' (*Plate 17*) was visible, with only a few exceptions, in the oviducts of all impregnated females and he believed that its functions were to prevent the loss of sperms and the entry of seawater. Hartnoll (1969) reported the production of a 'sperm plug' in several species of Brachyuran crabs including the European shore crab (*Carcinus maenas*) and the American blue crab (*Callinectes sapidus*). Modern workers agree with Williamson (1904) that the main purpose of the plug is to help pre-

Plate 16   Genital openings of a soft-shelled female crab with plugs visible

59

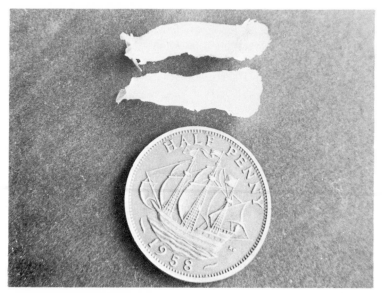

Plate 17   Plugs removed from the oviduct of an impregnated female. The coin is 25 mm in diameter

vent the loss of sperm after copulation. This is particularly true of species such as *Cancer pagurus* which possess a simple type of vagina and which mate when soft shelled. In these crabs the vulva would gape after copulation and so the presence of a sperm plug helps to reduce sperm loss. In *Cancer pagurus* the sperm plugs remain visible in the oviducts for some time, slowly disappearing as the shell gradually becomes hard. Observations by the author suggest that the plugs are usually visible in the entrance of the oviducts for three to eight weeks after copulation. Williamson reported that eventually the plug disintegrates. The author found remains of plugs, in a softened condition, in the spermathecae of females which showed signs of not having moulted for several years.

The presence of a sperm plug therefore appears to be a reliable indication that a female crab has been impregnated and this was used by the author to check on the size at which copulation occurred in different areas of the coast. This could be used as another guide to the size at which female crabs become sexually mature. Observations were first made during August 1961 on soft-shelled females during the main moulting season in the Yorkshire fishery. This was done while working aboard commercial crab boats, when all newly-moulted females larger than 100 mm carapace width found in the pots were

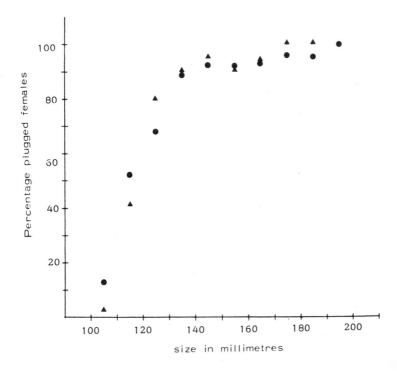

▲ YORKSHIRE - 1961

● SOUTH WEST IRELAND - 1968 and 1969

Fig 11   Percentages of soft-shelled females in different size groups found with sperm-
plugs – Yorkshire, 1961 and southwest Ireland, 1968–1969

examined for the presence or absence of plugs. Of the 331 soft females
examined, 75 per cent were plugged, those which remained unplugged
being mostly among the smaller crabs examined (*ie* 100–129 mm
width). The smallest female found with sperm plugs had a shell width
of 107 mm, but out of 37 crabs in the carapace width group
110–119 mm, 41 per cent were impregnated (*Fig 11*). Observations in
Irish waters show a very similar result (*Table 9*) and the proportion of
females with plugs increases in the larger size groups.

   Bearing in mind the fact that the author and other workers studying
Cancer pagurus have only rarely reported berried or hatching females
below 140 mm, it is surprising to record such a high proportion of im-

61

Table 9 Numbers and proportion of soft-shelled females in different size groups found with sperm plugs

Southwest Ireland

Width (mm)	90–99	100–109	110–119	120–129	130–139	140–149	150–159	160–169	170–179	180–189	190–199	Totals
Number examined	23	23	34	38	27	28	41	31	27	21	6	299
Number plugged	0	3	18	26	24	26	38	29	26	20	6	216
Percentage plugged	0	13	53	68	89	93	93	94	96	95	100	72
Yorkshire												
Number examined	0	37	46	47	42	44	44	37	24	10	—	331
Number plugged	0	1	19	38	38	42	40	35	24	10	—	247
Percentage plugged	0	3	41	81	91	96	91	95	100	100	—	74

pregnated crabs below this size. Hartnoll (1969) considers a crab as becoming mature when it enters the intermoult stage, during which it is first able to copulate successfully. However, he pointed out that in several species of crabs ovulation sometimes occurs a considerable time after copulation.

The author, in an attempt to explain impregnation in small females held 20 small (115–127 mm) crabs which were seen to be plugged in the laboratory tanks. After 12 months in the laboratory none of the crabs had become egg-bearing and no development of the gonads was seen to have taken place when the crabs were later dissected. These observations suggest that in *Cancer pagurus* mating can take place without subsequent egg-carrying, particularly in the smaller sized crabs. No other references have been found to verify this conclusion.

State of gonads. Williamson (1900) used the presence of fully developed gonads as an additional guide to maturity, although Hartnoll (1969) stated that, in Brachyuran crabs, the ovaries can still be immature at the time of copulation and that maturity cannot accurately be determined by the condition of the gonads. Even so it can give a guide to size at first maturity and in 1969 the author examined a large sample of Irish crabs for gonad development. The size groups ranged from 115 mm upwards.

Four maturity stages were recognised, which may be classified as:

Stage I   – No gonad development.
Stage II – Partial gonad development.
Stage III – Gonads extending into carapace. Orange in colour.
Stage IV – Ripe, carapace full of bright red gonad material.

These observations were made in August and September 1969 because a high proportion of female crabs have ripe, fully developed gonads at this time of year (*see later section on spawning*).

Studies showed that only 13 per cent of the crabs between 115 and 126 mm (4.5–5 inches) had ripe gonads (Stage IV) the stage which suggests that spawning will occur within a short time. However, in the larger size groups, 127 mm upwards, the proportion of ripe females ranged from 50 to 91 per cent (*Table 10*).

Table 10 Numbers of female crabs with developed gonads in various size groups – southwest Ireland, Aug/Sep 1969

			Stage			
Carapace width group (mm)	I	II	III	IV Ripe	Total in each size group	% Ripe
115–126	32	—	1	5	38	13
127–139	18	6	12	36	72	50
140–151	9	4	16	74	103	72
152–164	2	3	14	85	104	82
165–177	3	—	4	69	76	91
178–190+	3	3	4	72	82	88
Totals	67	16	51	341	475	72

These investigations on the size at which female crabs become impregnated in Yorkshire and southwest Ireland show that there is little difference in the data collected from both areas. Sperm plugs were present in 50 per cent of the crabs of carapace widths 108–115 mm (4.25–4.5 inches) indicating that copulation does occur in the smaller-sized crabs. However, observations on egg-carrying and hatching stocks suggest that in both areas a high proportion of the females do not carry eggs until they reach a carapace width of 127 mm and over.

Studies on the gonad development in various size ranges also suggest that a higher proportion of females in the large size groups develop full gonads.

Mating behaviour in *Cancer pagurus*

The published information in mating behaviour in *Cancer pagurus* is limited. Williamson (1900, 1904) presented some incomplete descriptions of copulation in the species but stated that the act of impregnation was not easily studied and that he had never personally witnessed copulation in *Cancer pagurus* either in the laboratory or on the shores. Pearson (1908) was one of the first workers to report that copulation took place immediately after the female had moulted, but he, like the other workers, never actually witnessed the act of copulation in *Cancer pagurus*.

During the period 1958–65 the author made a series of observations on the behaviour of crabs in tanks. These included a study on mating, including the act of copulation which was seen and photographed. This new information on mating behaviour added considerably to that presented by earlier workers and is described here in detail.

Pairing

During laboratory experiments it was noticed that a powerful attraction appeared to exist between male and female crabs prior to the female casting her shell. Although mating cannot take place until immediately after the female has completed her moult, when her shell is soft, pairing was found to occur for periods ranging from 3 to 20 days before the moult and for a further period of between 1 and 12 days after the moult (*Table 11* and *Fig 12*).

While in attendance the male crab sat astride the female (*Plate 18*)

Table 11 Carapace widths of male and female crabs taking part in pairing and the number of days of attendance by males before and after the female moult

(a) Female carapace width (mm)
(b) Carapace width of male attending at time of female moult (mm)
(c) Date of moult
(d) Attending period (days)
(e) Plugged

Table 11 cont.

(a)		(b)	(c)	(d)		(e)
Pre-moult	*Post-moult*			*Pre-moult*	*Post-moult*	
113	124	131	1. 8.59	12	4	No
113	137	135	10.10.59	11	12	Yes
120	132	130	13.10.59	6	7	Yes
117	132	131	17. 5.61	13	7	No
133	153	133	12. 9.61	5	2	Yes
93	115	107	18.10.61	13	3	Yes
94	118	102	26.10.61	3	2	No
120	139	133	1.11.61	21	3	Yes
104	123	107	3. 1.62	20	9	Yes
94	108	102	26. 4.62	20	9	Yes
119	125	109	17. 5.62	15	6	Yes
132	150	133	21. 9.62	3	4	Yes
105	127	127	22.10.62	4	5	Yes
113	135	125	24.10.62	9	5	No
94	118	108	26.10.62	3	2	No
91	113	127	8.11.62	3	1	Yes
109	131	127	16.11.62	10	2	Yes
93	115	131	14.12.62	9	3	Yes
116	141	115	16.12.62	10	4	Yes
137	157	131	12. 7.63	4	5	Yes
105	123	110	29. 7.63	4	3	Yes
107	125	110	13. 8.63	8	7	Yes
133	151	110	19. 8.63	3	5	Yes
111	132	113	13. 9.63	10	4	Yes
105	126	110	13. 9.63	8	5	Yes
115	136	113	18. 9.63	4	4	Yes
97	117	110	23. 9.63	5	3	Yes
111	132	131	2.10.63	7	8	Yes
111	132	131	2.10.63	8	6	Yes
105	124	110	3.10.63	8	5	Yes
106	127	111	8.10.63	11	4	No
92	115	110	11.10.63	3	5	No
111	132	110	14.10.63	7	5	Yes
122	141	110	22.10.63	4	6	Yes
114	136	118	18.11.63	5	1	Yes
109	132	120	19.11.63	4	7	Yes
105	125	110	25.11.63	9	3	Yes

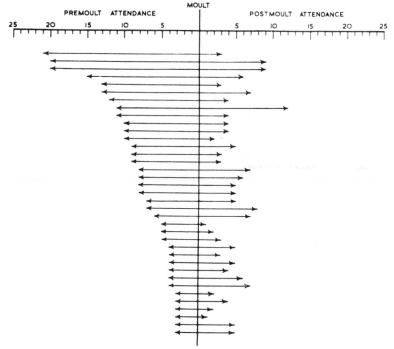

Fig 12   Number of days of attendance by male of *Cancer pagurus* before and after females moulted

with his claws held forward in a protective position, the pair remaining quietly in a secluded part of the tank.

Once attracted to a female the male showed no response to food and would remain continuously with the female, guarding her from the advances of other males in the tank. When disturbed, or if other males moved close to the female, the attending male became aggressive and sometimes attacked the intruder. The observation also showed that female crabs co-operated in this behaviour and they have on occasions been seen to actively help the male into position; also after separation, a female would sometimes herself return to the male.

In most cases the female crab was accompanied by the same male throughout the period of attendance but in a few cases a change in partner occurred during the early pre-moult attendance. Whenever there was a change a larger male took the place of a smaller one, which was sometimes seen to be actively displaced. The smallest

Plate 18   Male in attending position astride female

male which attended a female had a carapace width of 102 mm (4 inches), and copulation had apparently been successful since the female was later found to be impregnated, *ie* sperm had been introduced into her. In the laboratory tanks males tended to be polygamous; one attended and impregnated 3 females in 18 days and another (carapace width 110 mm) was observed to attend 10 different females during a period of four months. During this time there were 7 other males present in the tank yet this individual male paired with 10 out of the 17 females which moulted, and on no occasion was he replaced by another male. Polygamy by other crab species has previously been recorded for *Cancer magister* and for the king crab (*Paralithodes camchatica*) and the US blue crab (*Callinectes sapidus*).

In the author's experiments the majority of females examined during the first few days of attendance by males failed to reveal any signs of a forthcoming moult, although in some cases where a limb was missing a regenerating bud would be seen. The presence of a bud of a new limb has been found to be a reliable indication that the crab is about to cast its shell, the bud enlarging to a recognisable appendage after the moult. Another indication is that a few days prior to the moult the pleural line, which divides the tergal region of the carapace from the rest of the exoskeleton becomes more distinct and then at the moult it splits open (*see later section*).

Carlisle (Carlisle and Knowles, 1959) was of the opinion that the attraction between the sexes is chemical and states that ... 'For some time before the moult of copulation the female is exuding a substance which is attractive to the male ... the attractive substance can be detected by the male at a distance ... '. But he added that nothing is known about the control behaviour and the production of the attractive substance (pheromone) by the female.

This attraction observed to occur between crabs in the laboratory has also been observed on the foreshores off certain coasts of England by the author. Pairs of crabs are often seen on the shores, usually hiding under rocks near low-water mark, during July and August, the main moulting period on the east coast of England. These observations on the mating attraction have helped explain the temporary decline in the number of male crabs in commercial catches during part of the summer, when the males attend and remain with the moulting females, and are not attracted by baited traps.

Copulation

Williamson (1904) kept crabs in aquaria, and although he did not actually observe copulation he concluded that impregnation occurred immediately after the female had moulted.

During experiments at the Fisheries Laboratory, Burnham-on-Crouch, copulation between crabs was observed on several occasions, and each time it occurred between a hard-shelled male and a recently moulted female. In the majority of cases impregnation was known to have occurred within a few hours of moulting; in two cases, however, it did not occur until 48 hours after the moult. It does seem likely, therefore, that for successful copulation and impregnation the female must be in a very soft-shelled condition.

On one occasion the whole sequence of moulting and copulation was observed. During the moult the male remained in the pre-moult position astride the female, but supporting his own weight on his walking legs as though to avoid hindering the female's activities. At times the male appeared to be actually assisting the female with her moult by pushing off the carapace with his chelae (*Plate 19*). This behaviour was also observed between other pairs. Immediately after completing her moult, the female lay as an inert mass on the tank bottom, the male being alongside her. The male then gently turned the female over onto her back using his chelae to unfold the abdomen and

Plate 19   A male assists a female during a moult

expose the genital openings. Copulation then took place (*Plate 20*), in this instance lasting for three hours. On four other occasions, when copulation was observed in less detail, it was seen to continue for at least two hours, but in the majority of cases copulation was not observed because it took place at night. During copulation the male treated the soft female with extreme care and there was every indication that she co-operated with him during the act.

After the act of copulation, the sperm plugs, described earlier, become visible in the female vulvae. Four females were examined immediately after the male had completed copulation, and in each case the plugs were found in the oviducts. Copulation was observed to occur only once for each female, probably because the presence of the plug in the oviduct would prevent re-insertion of the male copulatory organ.

Hartnoll (1969) reviewed in detail a series of papers, including the author's published work, on courtship and copulation in nine species of crabs. For the *Cancridae* he reported that mating is basically similar and in five observed species there was evidence that males could distinguish pre-moult females. A post-moult attendance was also reported, and all accounts emphasised that the male treated the

69

Plate 20 A male astride a newly moulted female in the copulatory position

female gently and that the female was co-operative.

Spawning

A considerable amount of information is available on the spawning of *Cancer pagurus* (Williamson, 1900, 1904; Meek, 1904; Pearson, 1908). A review of the spawning cycle, based on the observations made by these workers, is summarised here. In addition, information collected by the author in Yorkshire and Ireland is presented, together with studies on egg attachment, which involved the use of modern high-powered microscopes not available to earlier workers.

After impregnation the paired female gonads commence development and extend in size until the ripe ovary covers almost the whole dorsal surface of the thoracic regions of the body (Williamson, 1900). Observations made by the author on gonad development in Yorkshire and Ireland confirm Williamson's studies and can be summarised by the following stages:

Ovary immature—gonads small and pale yellow in colour, sometimes with a slight greenish tint. Eggs small, average diameter of 0.1 mm; not yolked.

Ovary developing—gonads extending into carapace, orange or light red in colour, eggs yolked, about 0.2 mm diameter.

Ovary ripe—gonads extending, cover dorsal body, bright red in colour. Eggs can be seen with naked eye, yolked, with a diameter of 0.4 to 0.7 mm.

Examination of soft-shelled crabs showed that the ovary is not ripe at the time of impregnation. The ovary may not be ripe for a further 3 months after copulation or in some cases up to 15 months later (Williamson, 1900). Pearson (1908) also agreed that in many cases spawning did not follow in the same year as impregnation occurred, but was delayed until the following winter, *ie* 14–15 months later.

In November 1964 the author, working in Yorkshire, attempted to obtain precise data on the time-gap between impregnation and spawning. This was determined by examining the gonads of crabs known to have been impregnated, and which had therefore moulted in the summer of 1964. At the same time a random sample of females, which had not moulted that summer, was also examined (*Table 12*).

Table 12 Numbers and percentage of crabs in various size groups and shell conditions with ripe gonads – Yorkshire 1964

Size group (mm)	Recently moulted			Not recently moulted		
	No undeveloped	No ripe	% Ripe	No undeveloped	No ripe	% Ripe
115–139	38	—	Nil	9	8	47
140–164	23	25	52	4	13	76
165–180	6	30	83	5	16	76

These results suggest a tendency for spawning to follow within a few months of impregnation, particularly for larger females. Thus, 83 per cent of the recently moulted females in the size range 165–180 mm had ripe gonads and were ready to spawn; only 52 per cent of the females in the middle size group (140–164 mm) were ripe, whilst there were no smaller crabs (115–139 mm) with fully developed gonads. It is worth noting that all the crabs which had recently moulted were known to be impregnated because of the presence of sperm plugs. In the case of the unmoulted females it was not possible to determine whether they had moulted and been impregnated.

However, the higher proportion of ripe crabs in the smaller size groups may be a result of delayed spawning. This phenomenon has been observed in *Cancer pagurus,* and following impregnation, two or three successive batches of eggs may be produced without an intervening moult (Williamson, 1900; Pearson, 1908). This means, therefore, that one supply of spermatozoa is probably sufficient to fertilise several egg batches. Continuous spawning has also been recorded by the author at Burnham; five crabs, held for periods ranging from 16 to 28 months, spawned twice without moulting, but none ever spawned three times in a row.

Spawning commences along the east coast of Britain in late November and continues until February. At this time the females are offshore in the deeper water, where they spend the winter (Williamson, 1904; Meek, 1904; Pearson, 1908). Spawning is also reported to occur off western Norway between the months of October and January (Nordgaard, 1912).

Studies by the author on the gonad development of the Yorkshire stocks showed that ripe females were not present in the catches during the spring or early summer. However, the numbers increased in the later summer and by November over 70 per cent of the females over 140 mm had fully developed gonads (*Table 13*), suggesting that spawning was imminent.

Table 13 Monthly observations on gonad development – Yorkshire 1963 (Total samples examined from 200–250 crabs per month)

Months	J	F	M	A	M	J	J	A	S	O	N	D
Percentage with fully ripe gonads	NS	NS	0	0	0	0	0	3	10	53	71	22

NS = No sample

In Ireland greater emphasis was placed on investigating the seasonal variation in the development of the gonads, because of a market demand from Sweden for ripe females. Regular monthly samples were taken by the author and examined to determine gonad development. These data from the southwest of Ireland are shown in *Fig 13*. A few ripe females were taken early in the summer but the

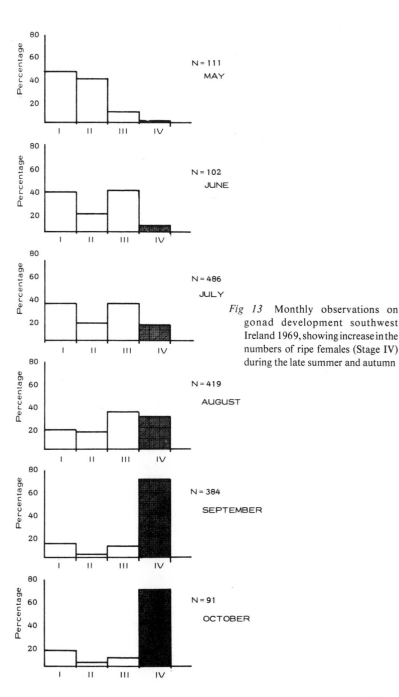

Fig 13 Monthly observations on gonad development southwest Ireland 1969, showing increase in the numbers of ripe females (Stage IV) during the late summer and autumn

proportions increased in later months. Based on these observations it was concluded that spawning in southwest Ireland commences in late October, *ie* about a month earlier than in the east of Britain. This conclusion is also borne out by the fact that commercial catches in this area fall to a very low level in October, although further west – in the Galway and Sligo area – development is later and spawning appears to occur from November onwards. Exploratory fishing and gear trials in October and early November 1967 showed that on some sandy grounds in Roaring Water Bay, Co. Cork, the catch was predominantly of females, most of which had ripe gonads. It was concluded that this stock had congregated in this area for spawning.

Laboratory observations on spawning
During the period 1961–64, 34 records of spawning were made by the author in the laboratory tanks at Burnham. These observations showed that even in captivity the main spawning period was in the winter months; for example, 16 crabs spawned in January, 13 in December, one in November and one in March. The remaining three spawned in August.

Laboratory experiments also showed that, for successful egg attachment, females must spawn on a soft bottom. Early in these investigations it was observed that if a crab spawned in a concrete tank very few eggs became attached to the endopodite setae but, instead, remained loose on the base of the tank. Williamson (1904) had reported that spawning females selected particular areas off the Scottish coast where the sea bed was composed of sand or shingle.

In December 1963 a small experiment was arranged by the author to test whether females would select a particular type of bottom in the laboratory. For this test the bottom of a concrete tank one metre square was therefore sub-divided into two sections, one being filled with a sandy type of gravel and the other left bare except for a few small rocks which were covered with seaweed. Ten large females (150–170 mm carapace width) which had not moulted that summer and ten males of similar size were collected from Yorkshire and established in the tanks. Bearing in mind the observations on gonad development made earlier (*Table 13*) there was every likelihood that many of these females would be ripe.

The procedure adopted for a 10 day period was to place all the crabs onto one substrate at 1700 hours and to record their positions the following morning at 0900 hours. That evening the crabs would be placed on the other bottom and their positions similarly recorded the following day. This alternating positioning procedure was continued for the whole period of the experiment.

The results were most conclusive. On each of the ten mornings the females were *all* found on the shingle, whereas the males had dispersed themselves throughout the tank – some being on the shingle and some under the rocks. Later that month and during January 1964, five crabs spawned in the shingle area, and the eggs became attached to form the usual egg mass found on crabs (*Plate 15*) which spawned in the field.

On one occasion spawning was observed. The female became partially buried in the sand, and during the period of egg-laying, which lasted for about four hours, this crab continually moved her abdomen in an upwards and downwards motion. It appears from the behaviour observed that the female digs out a small hollow into the sand, and this forms a collecting pouch between the body and the curved abdomen into which the eggs are extruded. The need for a seabed for spawning composed of sand or shingle therefore helps to explain why stocks of ripe and berried females are regularly found in sandy areas in the late autumn and winter months.

Egg attachment and development
The process of attachment of eggs to certain setae of the pleopods was studied in *Cancer pagurus* following spawning in the laboratory tanks. The opportunity of examining newly spawned eggs, using a Zeiss Photomicroscope with magnifications up to 2,000, enabled some new information to be collected.

In the first instance eggs from crabs spawning on a concrete base, which failed to attach to the pleopodal setae, were examined; these eggs were found to have a space between the membrane and the yolk (*Plate 21*). Williamson (1904) described this as a perivitelline space and he concluded that it contained a fluid possessing adhesive properties. Tests were made at Burnham to check whether these recently spawned eggs stuck to each other. Newly-laid, but non-attached, eggs collected from the base of the tank were left in a beaker of sea water; none stuck together, suggesting that they were not sticky when laid. However, it was found that, if they were left overnight, on the following day they were massed together. This observation agrees with the studies made by Burkenroad (1947) on the eggs of *Palaemonetes,* which were not adhesive when laid but became

Plate 21   A loose egg taken from the base of a concrete tank (note large vitelline membrane)

adherent to each other later.

Microscopic examination of eggs attached to the endopodite setae suggested that there is initially a cement bond between the egg membrane and the hair (*Plate 22*) but eventually this outer membrane is pulled out, probably by the weight of the egg itself, to form a funiculus or stalk. *Plate 23* shows the final stage of egg attachment with the eggs attached by stalks to the setae.

Williamson (1904) was perhaps one of the earliest workers who dealt with the subject of egg attachment in the *Brachyura*. However, his suggestion that each egg was impaled on a hair which had been thrust through the external egg membrane was difficult to accept. Broekhuysen (1936), studying *Carcinus maenas,* slightly modified the impaling theory by assuming that the egg was passed onto the hair instead of being impaled against it. One of the most comprehensive studies on egg attachment was made by Yonge (1955) for the lobster *Homarus*. These investigations suggested that the eggs were covered with cement at the time of spawning so that they were glued to the setae, which also formed the outer membrane or egg capsule (funiculus). Cheung (1966) reviewed in considerable detail the development of egg membranes and egg attachment in *Carcinus maenas, Homarus vulgaris* (now *H. gammarus*), *Nephrops norvegicus*

76

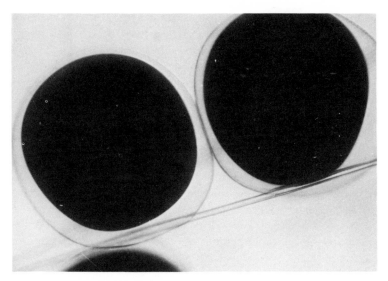

Plate 22   Eggs after 12 hours of becoming attached to an endopodite setae of a berried female

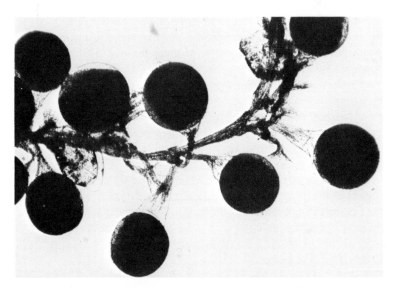

Plate 23   Eggs after 36 hours of becoming attached to the endopodite setae of a berried crab

77

and *Astacus pallipes*. This author found that no fundamental difference existed in the membranes of these species, and he concluded that the funiculus for attachment is formed from the outermost layer of the egg membrane. In the same paper the author summarised the earlier studies on egg attachment in Decapod Crustacea.

In general the main controversial elements of previous work concern the origin of the glue or cement and the process of egg attachment. In species such as *Homarus* and *Astacus* some workers considered that the cement originated in the oviducts, while others believed that the tegumental glands on the swimmerets produced the adhesive secretions. It must be remembered, however, that an important difference between the *Macrura* and the *Brachyura* is that the former have external spermathecae, while the latter have internal ones. Cheung (1966) could find no evidence of cement glands in the oviducts of *Carcinus* but he believed that the glue was liberated from the egg itself by being squeezed. This process may also apply in *Cancer pagurus* because it has been observed that the perivitelline space, which is large in loose eggs, reduces in size after attachment (*Plate 22, 23*).

Newly-spawned eggs of *Cancer pagurus* are orange in colour but they gradually become redder as development continues. When ready to hatch the eggs are a dirty-grey colour and the black eye-spot of the larvae is visible. Measurements of the egg diameter range from 0.37 to 0.41 mm. Lebour (1928) reported a diameter of 0.32 mm early in development and 0.4 mm prior to spawning. Unfortunately studies on the development of the eggs on crabs held in the tanks at Burnham were limited because in most cases the females lost their eggs within a month or two of spawning. Measurements of the egg diameter were therefore limited to a short period, but observations on berried crabs caught at sea showed that even in the spring (prior to hatching) the eggs rarely exceeded a diameter of 0.5 mm.

Why this loss of eggs occurred in the laboratory tanks is not known, but it may have been due to the salinity fluctuations which were known to occur in the estuary of the River Crouch, from where the tank water was pumped.

Numbers of eggs carried
Buckland (1877) stated, as a result of a calculation made by his secretary, that the crab carries no fewer than 1,441,000 eggs; Williamson (1900) presented data on the number of eggs carried externally by

six females of varying sizes. He calculated the number of eggs carried by the following method: 'The swimmerets with the attaching eggs were preserved in spirit. They were then dried in a water bath and weighed. From the weight thus obtained was subtracted the weight of the dried swimmerets of a crab of the same size, but which was not berried. The weight of the eggs was thus found approximately. The weight of the hardened cement by which the eggs were attached to the swimmerets was neglected. The number in a small weight of the eggs was counted, and from that datum the total was deducted.'

Williamson's results are shown in *Table 14* and he concluded that there appeared to be an increase in the number of eggs carried with an increase in the size of crabs.

Further information was obtained by the author when counts of the numbers of eggs carried were made on 10 berried crabs caught off Yorkshire in 1965. The method used differed slightly from that used by Williamson but it was still time-consuming; hence the small number examined. The eggs and swimmerets were first preserved in

Table 14 The number of eggs carried externally by berried crabs of various carapace widths

(a) After Edwards (1967)

Shell width (*inches*)	Number of eggs carried	Shell width (*inches*)	Number of eggs carried
5.7	1,632,000	6.8	2,524,000
6.0	823,800	6.8	2,451,000
6.3	1,847,000	7.0	2,176,500
6.8	1,766,800	7.2	2,876,500
6.8	1,995,500	7.4	2,623,500

(b) After Williamson (1900)

Shell width (*inches*)	Number of eggs carried
5	1,010,000
5.75	460,000
5.75	750,000
6	940,000
6	1,480,000
7	3,000,000

Bouin's fluid for 14 days to harden the eggs. After drying in an oven for 10 days at 100°C the eggs were easily removed from the swimmerets, although many of the fine hairs were also collected. It was, however, possible to remove these by sieving and the clean eggs from each crab were then weighed. Smaller sub-samples of about 1–2 g were then taken and counted, and from these the total number of eggs carried was calculated. The results are shown in *Table 14* along with the data presented by Williamson (1900). There appears to be a tendency for the larger crabs to carry more eggs, although some individual variations occurred.

Hatching and larvae
Williamson (1900), Meek (1904) and Pearson (1908) all recorded that the eggs remained attached to the female's endopodite setae for a period of seven to eight months. According to these workers, hatching occurred in the inshore waters during the late spring and continued during the summer. Nordgaard (1912) also reported an inshore migration for hatching in Norwegian waters during July and August.

Larvae were found in Danish waters from April to late October (Thorsen, 1950) while Rees (1952) records larvae in the North Sea from May to December with peaks occurring in July, August and September. In the inshore plankton off Plymouth, Lebour (1947) took crab larvae from April to September and again in November. The greatest numbers, however, occurred in May and July. According to Williamson (1956), *Cancer pagurus* larvae were abundant in the Irish Sea in May, September and October. Extensive plankton studies by O'Ceidigh (1962) in Irish waters showed that on both the east and west coasts of the country the various larval stages of *Cancer pagurus* were common in the inshore plankton from March to September.

No investigations on the larval distribution of crabs were made by this author. However, observations on the occurrence of hatching female crabs were made, both in England and Ireland, to establish the period when the larvae were released. A description of how recently hatched females can be identified has been given earlier in this account. Observations by the author on the occurrence of hatching crabs in the commercial samples examined in Yorkshire and south-west Ireland are shown in *Fig 14.* and agree with the findings of the earlier workers mentioned above.

Observations off Yorkshire in 1963 showed that hatching appeared to commence in April and reached a peak in August (*Fig 14*). In

80

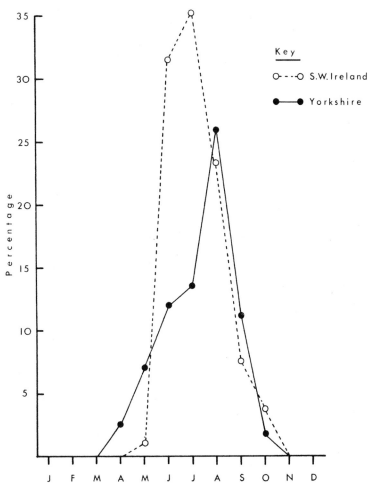

Fig 14 Proportions of hatching females in the catches. Yorkshire 1963 and southwest Ireland 1969 and 1970 ·

Ireland during 1969 and 1970 hatching crabs were abundant in June and July, but the numbers declined from August onwards. The peak hatching time in Ireland appears to occur slightly before that observed in Yorkshire waters, but the results show, however, that the main hatching period in both areas is from May to September.

The larval stages of *Cancer pagurus* have been well described by Williamson (1911), Lebour (1928) and Gurney (1942). No additional

81

investigations on this aspect of the crab's life-history appeared to be necessary. According to Williamson (1911) and Lebour (1928), after a short protozoea stage, development continues through five zoea stages before the megalopa stage is reached. By this time the larval crab has a size of 2.24 mm (Lebour, 1928); within a day it settles to the seabed and assumes the first crab stage (*Fig 15*). The larval period, which lasts for 23–30 days, is a dispersal phase because during it the larvae can be transported considerable distances from where they are first hatched (Lebour, 1928).

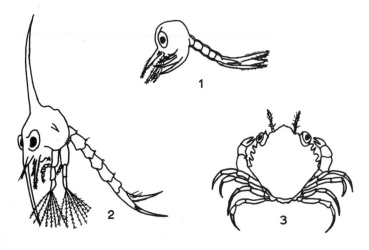

Fig 15   Stages in the development of the young crab. *1* Protozoea, *2* Zoea, *3* first post larval stage (after Lebour, 1928)

5 Growth

General

In crustacea such as crabs, lobsters and shrimps the body is completely enclosed in an exo-skeleton (outer shell) of calcified chitin. Because it is enclosed in this hard rigid shell growth is not continuous and only occurs when the shell is cast. This process is known as moulting or ecdysis. Growth in crustacea is therefore a combination of the increase in size at a moult (moult increment) and the number of times moulting occurs (moult frequency).

The nature of crab growth by moulting had led to much difficulty in determining the true growth rate under natural conditions, since tags or marks attached to the exo-skeleton are lost when the shell is cast. Almost all previous observations on the increase in size at moulting in *Cancer pagurus* have therefore been made on crabs kept in tanks (Williamson, 1900, 1904; Meek, 1904; Pearson, 1908), and these figures may not give a true guide to growth under natural conditions. However, in 1957 a new type of tag known as the 'suture' tag, which is not lost when the crab moults, was developed by W A Van Engel of the Virginia Fisheries Laboratory, USA for tagging the blue crab (*Callinectes sapidus*). This tag, after certain modifications to the tagging techniques, was found suitable for use on *Cancer pagurus* to obtain moult increment data (Mistakidis, 1959, Edwards 1964).

This suture tag was further tested by the author in the laboratory tanks at Burnham and in 1959 and 1960 in field experiments in the Norfolk crab fishery. The results from these preliminary experiments were encouraging and the tag was found to be a practical and effective method for obtaining information on moult increments under natural conditions. In addition it was found that some tags were retained by crabs through several moults and this added further to the information on the annual growth and also on the migrations of *Cancer pagurus*.

This chapter describes the moulting process in *Cancer pagurus*, the soft-shelled condition and the results of growth experiments made by the author.

The moulting process

In the last two decades much work has been published on moulting and its control in Decapod Crustacea. Carlisle and Knowles (1959) in their book 'Endocrine Control in Crustaceans' review most of the literature, including the classical work of Carlisle and Dohrn (1953). Other important studies are those by Drach (1939) and Passano in Waterman, ed., (1960).

The term moulting is used here to describe the act of withdrawing from the old integument and the state of post-ecdysis, including the increase in linear size. The importance of the moulting process in relation to the crab's metabolism, reproduction and behaviour must, however, be taken into account, but is not considered here.

Various authors have described moulting in *Cancer pagurus*. According to Williamson (1900), a few days prior to the moult the crab retires to a retreat, usually among rocks, which affords some protection. The fact that decapods typically shed their exo-skeleton at a time and a place of their own 'choosing' strongly suggests that the initiation of moulting is, in some unknown manner, under nervous as well as hormonal control (Waterman, ed., 1960). The descriptions of moulting published by Williamson (1900) and Pearson (1908) are rather limited; however, during the growth studies made at Burnham-on-Crouch, when 120 crabs moulted in the laboratory tanks, the author witnessed the moulting in *Cancer pagurus* on several occasions and was able to photograph the sequence (*Plates 24–26*). A brief description of a typical moult, as witnessed in the laboratory, is given below.

Plate 24   The moult commences and a split develops along the pleural line

Plate 25   The crab slowly draws its abdomen from the old exoskeleton

Plate 26   The moult is completed. The enlarged soft-shelled crab lies alongside its exoskeleton

The moulting crab usually moved to a secluded part of the tank and within a few days a split developed between the epimera and the inturned edge of the carapace along the pleural groove. The split gradually enlarged until the carapace was freed from the abdominal part of the exo-skeleton.

After a short period of muscular body movements the soft crab slowly backed out of the old integument, the abdomen being withdrawn first, followed by the walking legs, each being freed individually. These periods of straining were followed by periods of rest, when the crab remained completely quiescent. Once all the eight walking legs were free they were used as levers to assist in extracting the chelae, which were always the last appendages to be withdrawn from the old exo-skeleton. Immediately after the moult was completed the crab lay by the cast exo-skeleton in a soft blubber-like condition, and usually remained exhausted for about an hour. Examination of the cast exo-skeleton showed that this consisted of all the limbs and appendages, including the gills, mouthparts, antennae and eyestalks. In fact so complete is this exo-skeleton that on several occasions the author mistook it for a dead crab.

The studies in the Burnham laboratory tanks showed that the actual period of ecdysis ranged from 30 minutes to six hours and the process usually occurred at night. However, if moulting was abnormally prolonged it usually ended in death. The only other time period recorded for this species is one published by Williamson (1940), who reported that a crab completed a moult in an aquarium in 20 minutes. In the laboratory tanks a moulting mortality of about five per cent occurred. Some crabs failed to extract themselves from their old exo-skelton and others were attacked by other crabs in the tanks. The protection given by males to moulting females, described earlier, helped to reduce the mortality of females by cannibalism, but several males died in this way.

Immediately following ecdysis the soft-shelled crab increases its size due to absorption of water, but within a very short period, usually within 24 hours, the new exo-skeleton commences to harden and the final linear dimension is reached. Measurements taken by the author (*Table 15*) on five recently moulted crabs show that the final size is usually attained within about four hours of completing the moult. No further change in linear dimensions will then occur until the next moult.

Even though no further size increase takes place the crabs are still

Table 15 Linear increases (mm) of *Cancer Pagurus* after moulting in laboratory tanks

	Crab 1	Crab 2	Crab 3	Crab 4	Crab 5
Time of Measurement	*Male*	*Female*	*Male*	*Female*	*Male*
			Pre-moult size (*mm*)		
	114.0	*108.0*	*86.0*	*105.5*	*81.0*
Immediately after moulting	125.5	127.0	96.0	120.0	90.8
1 hour after moulting	129.4	127.6	—	120.0	—
4 hours after moulting	131.0	128.0	100.5	120.0	—
24 hours after moulting	131.0	128.0	100.5	123.0	100.1
4 days after moulting	131.0	128.0	100.5	123.0	100.1
Final Increment (*mm*)	17.0	20.0	14.5	17.5	19.1

very soft-shelled for up to seven to eight days after moulting. However, in the tanks these crabs were observed to move about and feed, and in fact at this stage they are taken in commercial pots in large numbers and have to be rejected at sea because of the laws which protect them in the UK.

Hardening of the shell

According to both Williamson and Pearson, crabs which moult do not reach a final stage of calcification for at least two or possibly three months after ecdysis. Tank observations made by the author agree with these findings, but it is known that environmental and behavioural factors, such as low water temperatures and fasting, can prolong the time required for the shell to finally harden. Observations made by the author off the Yorkshire coast show that crabs which moulted in July or early August become hard-shelled before November of the same year.

Legal protection for soft-shelled crabs dates back to the last century. Under Section 8 of the *1877 Fisheries* (*Oysters, Crabs and Lobsters*) *Act* (part b) it was an offence to land or sell. '. . . Any edible crab which has recently cast its shell whether known as caster, white-footed crab, white-livered crab, soft crab or glass crab or by any other name.' Furthermore in certain areas of England additional local bye-laws were also introduced; for example, in the Northumberland Sea Fisheries Committee's area a closed season for crabs from October 1 to December 31 each year was introduced in 1915 (later changed to

1 November–31 December). The sole aim of this bye-law was to protect soft crabs from damage or from being used as bait for long-line fishing.

The hardening of the crab's exo-skeleton is progressive, and various stages such as 'glass' crabs and 'white-footed' crabs have been long known to fishermen are mentioned in the 1877 Act.

Laboratory and field observations on white-footed crabs were made by the author during the course of the east coast crab studies and the results are described below.

Studies on white-footed crabs

At the beginning of each fishing season (April–May) up to 20 per cent of the male crabs, but only a few females, caught on the Norfolk grounds are in the white-footed condition. These crabs can easily be recognised by the colour of their claw pincers, which are greyish white compared with the normal black colour, and by the speckled and clean appearance of their shells. The shells, although they appear hard, have not yet reached the final stage of calcification. The meat condition of white-footed crabs is generally poor and a local bye-law prohibits the landing of this type of crab until after 30 June, when their condition has improved. Before this date white-footed crabs are normally rejected at sea, and any which have been overlooked and included in the landed catches are returned to the sea by the local Fishery Officers.

Observations made by the author (Edwards 1966a) while accompanying Norfolk crabbing boats showed that white-footed crabs were abundant on the grounds from April to July, but from then on their numbers declined, only occasional specimens being seen in September and October. Around 80 per cent of the white-footed crabs were males, and this condition appears to occur in crabs which have not moulted until late autumn, with the result that their subsequent improvement in condition has been delayed until the next year, due to the reduction in feeding at the low winter sea temperatures. During recent surveys of other east coast crab fisheries only occasionally were white-footed crabs ever found, and the abundance of this type of crab on the Norfolk grounds is believed to be associated with the shallow depth of the coastal waters in this area.

Laboratory observations. For some years it has been the general belief amongst Norfolk fishermen that white-footed crabs remain permanently in this condition and never become normal. In May 1965, 50 crabs in the white-foot condition, supplied by two Sheringham fishermen, were established in laboratory tanks at Burnham-on-Crouch. By early August 70 per cent of them had become normal crabs with black pincers and fully hardened dark-coloured shells, while in the remainder the pincers had become nearly black. Regular examinations during the holding period showed that the claw pincers gradually changed in colour from white to light grey, and then to dark grey and black. In the Norfolk fishery crabs in the intermediate stage, with grey/black pincers, are known as 'turntoe' crabs.

Meat yield observations showed that at the start of the experiment in May the meat in these white-footed crabs was watery and low in yield. However, in August, although the crabs had been held under artificial conditions, the meat was less watery and the yield had improved. It is to be expected that this improvement in condition would be greater under natural conditions, and since white-footed crabs are accepted by merchants after 30 June, their meat yield must then be reasonable. This experiment shows that white-footed crabs held in laboratory tanks do eventually assume a normal appearance and improve in condition.

Observations under natural conditions. In order to find out whether the change from the white-footed condition was a normal occurrence in the sea, 300 crabs classed as white-footed were claw-tagged early in June 1965 and released on the Norfolk inshore potting grounds. During the 1965 season, 186 (62 per cent) were recaptured and the colour of their claw pincers was recorded by the Fishery Officers; many of those still white-footed were returned to the sea and several were caught a second time. The records show that by late June many of the recaptured crabs were changing in condition and were in the intermediate turntoe stage, whilst in early July a few were being recaptured in the normal black condition. During late July and early August most of the tagged crabs caught were recorded as having either black or nearly black pincers, and only the tag identified them as having been previously in the white-footed condition. Recoveries did, however, show that even in September a few were still classed as white-footed, and the speed with which the improvement in condition takes place appears to vary with individual crabs (*Table 16*).

The results, nevertheless, clearly confirmed that the white-footed condition is only a temporary phase and appears to be caused by moulting late in the season when low sea temperatures reduce feeding. Furthermore, laboratory experiments in which moulted crabs were held at low temperatures (5–7°C) showed that shell hardening was slowed, and the crabs kept under these conditions became typical white-footed crabs.

Table 16 Proportion of white-footed crabs tagged and released on 9–10 June 1965 that were recaptured in the various stages during the following three months of fishing

Month (a) 1st part (b) 2nd part		Condition of crabs on recapture		
		Percentage 'white-footed'	Percentage 'turntoe'	Percentage 'black'
June	(a)	83	17	Nil
June	(b)	51	49	Nil
July	(a)	41	51	8
July	(b)	29	62	9
August	(a)	20	65	15
August	(b)	7	60	33
September	(a)	6	40	54
September	(b)	Nil	30	70

The moulting period

According to Williamson (1904) the main moulting period for *Cancer pagurus* on the east coast of Scotland is from July to September, but it may extend into December. In the Isle of Man moulting occurs from August to November (Pearson 1908). Neither author presented any substantial data to support his conclusions.

Newly moulted crabs can easily be recognised by the clean appearance and soft condition of the shell, which does not become completely hardened until two or three months after the moult. However, because soft-shelled crabs are protected by national regulations and not landed it is only possible to determine their abundance in the catches by accompanying commercial boats to the various grounds. Observations of this type were made by the author in the Norfolk fishery (Edwards 1966a), off Yorkshire (Edwards 1967), and around Ireland (Edwards and Meaney 1968).

Along the east coast of England the main moulting period was found to commence in July and to continue until October, although a few soft crabs were still found in November. *Fig 16* shows that in this area soft-shelled crabs were present in the catches during most months of the year but reached a peak in August, September and October, when they comprised between 30 and 50 per cent of the commercial

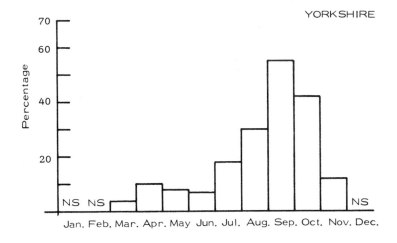

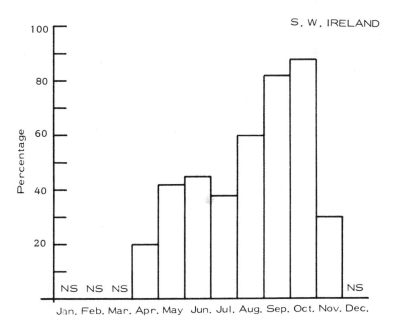

Fig 16 Proportion of soft-shelled crabs in the commercial catches, Yorkshire 1961 and 1962 and southwest Ireland 1968 and 1969 (NS = No sample)

91

catches. This is one of the reasons why many English boats discontinue crab fishing after July and concentrate on lobsters or other species.

Although the main moulting period off the east coast of England appears to be well defined there is some variation in the time of year when it starts. For example, in 1961, moulting commenced in early July, whilst in 1962 it occurred later in that month. Observations made in 1963 indicated that in that year moulting did not commence until August, and this delay is believed to have been associated with the severe winter of 1962/63 and the low sea temperatures which prevailed until May.

Studies in Yorkshire (Edwards 1966b, Edwards 1967) showed that during July and August the recently moulted crabs of commercial size (*ie* 115 mm and over) consisted of more females than males. In September and October the number of soft-shelled males had increased, and the shells of most of the females had hardened. Although there was some overlap, it appears that along the east coast of England there is a well defined moulting period and that females moult before males. This difference in timing is believed to be related to the mating act, which occurs between hard-shelled males and recently moulted females.

Preliminary observations in southwest Ireland (*Fig 17*) and in Cornwall (Bennett and Brown 1970) have so far failed to identify a well defined moulting period, and large numbers of soft-shelled crabs were found in both areas in the spring and also later in the year. There was also no indication that females moulted before the males. It appears, therefore, that the defined moulting period known to occur in Scotland and in other east coast crab fisheries may be in some way associated with the environmental conditions found in these waters.

Experiments to estimate growth in crabs using the suture tag

Bearing in mind that the nature of growth by moulting in *Cancer pagurus* has led to much difficulty in determining the true growth rate under natural conditions, the development of the suture tag, which is not lost when the crab moults, has been a most valuable advance and has made it possible to determine both the increase in size at moulting and the frequency of moulting. Studies using the suture tag have been made off Scotland by Mason (1965), off Norway by Gundersen (1963), in Swedish waters by Hallback (1969) and in south-west

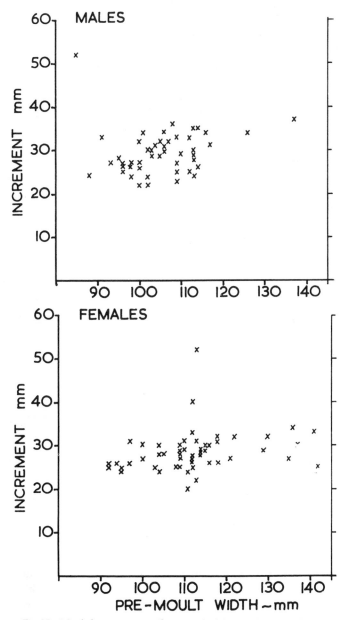

Fig 17 Moult increment data from moulted crabs released off Yorkshire in 1962 and 1963

England by Bennett (1970, 1974) and Bennett and Brown (1976). The author carried out a major tagging experiment along the east coast of England from 1959–66 when some 12,000 suture-tagged crabs were released in the fisheries off Norfolk, Yorkshire and Northumberland. This section of the book summarises the author's suture tagging experiments and briefly describes the technique which has proved a successful method of collecting growth data for *Cancer pagurus*.

The suture tagging technique
In the original method first used on *Cancer pagurus* in 1957 by Mistakidis (1959) stainless steel wire was used to attach the tag to the crab's body (*Plate 27*).

The original method involved the following sequence of actions:

(*1*) Two holes were pierced, 13 mm apart, directly on the line of separation of the carapace, just above the base of the walking leg. This was done with two bronze gramophone needles embedded in a 'perspex' holder, thus ensuring that the holes were a constant distance apart.

(*2*) A 15 cm length of monofilament stainless steel wire (30 swg) was passed through the holes by means of a curved surgical needle (Ferguson's $\frac{1}{2}$ circle No 10). This operation was simplified by using a MacPhail's Needle Holder to hold the needle during insertion.

(*3*) The ends of the wire were tied into a square (reef) knot, allowing sufficient space between knot and carapace for growth. The tag was then threaded on to the free end of the wire and secured by a further square knot. All the tags used in these experiments were of the Petersen button-type, which are ivorine plastic discs, 12.5 mm diameter and 1.5 mm thick.

During the initial experiments the above method was used successfully on crabs with carapace widths between 80 and 200 mm. The average time taken to tag each crab was two minutes, which included measuring, securing the tags and recording the data. Two workers, using this technique, could tag between 150 and 200 crabs each day. Tests also showed that the mortality caused by the tagging method was less than five per cent.

Using this method some 1,228 crabs were suture tagged and released off the Norfolk coast during 1959 and 1969; 439 (36 per cent) were later recaptured and of these 39 had moulted and increased in size,

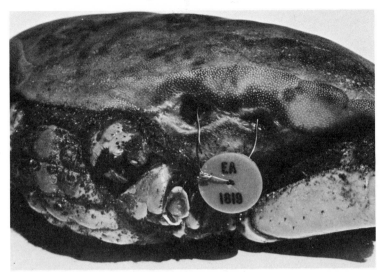

Plate 27 Suture tag attached to a crab by stainless steel wire, characteristic post-moulting discolouration around tag holes and enlarged tag holes

the tag remaining attached to the new exo-skeleton.

However, the original technique was not fully successful because the pierced holes became enlarged due to the stainless steel wire cutting the soft shell (*Plate 27*). Several fishermen also reported seeing crabs which showed visible signs of having been tagged (*ie* pierced holes) but no tag was present.

Laboratory experiments also showed that the wire did cut the shell, particularly when the crab was in the soft-shelled condition after moulting. On the basis of these observations it was concluded that some tag loss occurred under natural conditions, which would, of course, affect the results.

Modified technique. In an attempt to reduce this damage to the shell and the possible loss of the tag, trials were made in 1962 using a braided nylon thread instead of wire (*Plate 28*). The distance between the pierced holes was also increased from 13 to 23 mm. Experiments in the laboratory and the field confirmed that braided nylon was more suitable than the stainless steel wire previously used and it was used in experiments off Norfolk and Yorkshire during 1962 and 1963.

In 1965 a further minor modification was introduced when a lead seal clamped to the braided thread was used to replace the knots

95

Plate 28 Suture tag attached by braided nylon which is knotted. (Note the small tag holes, even after moulting)

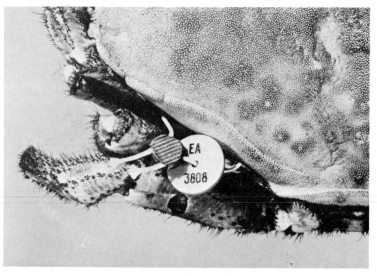

Plate 29 Suture tag attached by braided terylene which is clamped with a lead seal

(*Plate 29*). Braided terylene was also used instead of nylon since it is cheaper and has a better gripping power. This method has been used successfully by Bennett (1974) in studies on crab growth in southwest England and by the author off the east coast of England.

Results of growth experiments

During the years 1959–66 the author collected about 400 moult increment measurements from moulted suture tagged crabs in the fisheries along the east coast of England. This allowed estimates to be made of moult increments, moult frequency and annual growth rate. Bennett (1974) also collected similar data from crab fisheries off Devon, Cornwall and Dorset. Typical increments for crabs of both sexes are shown in *Fig 18* and *Table 17*.

Based on the results of suture tagging recaptures it appears that on the east and northeast coast of England the average increase in size after one moult was between 20 and 25 per cent of the original size. The average increment for females on both the Norfolk and Yorkshire fisheries was about 24 mm for crabs varying between 90 and 145 mm carapace width. In Yorkshire male crabs had a slightly larger moult increment than females (about 27 mm), but in Norfolk the available data for males suggest a slightly smaller moult increment than for females.

The lack of information on moult frequency has partly been overcome by means of tag recapture data, using the 'anniversary method' (Hancock and Edwards, 1967). This technique gives an estimate of moult frequency by reference to the proportion of crabs caught one year after release which have moulted. Males and females were considered separately, and it was found that annual growth in males becomes reduced at a smaller size than in females. The results from a series of suture-tagging experiments in the crab fisheries along the east coast of England show that annual growth in *Cancer* slows down in both sexes after a carapace width of 100 mm (4 inches) is reached. Furthermore, for males the proportion moulting annually decreases earlier than for females, and this results in a faster growth rate for females than for males of a similar size.

Estimation of growth in *Cancer pagurus* is further complicated by the reproductive cycle. Williamson (1940) has pointed out that the presence of sperms in the female spermatheca arrest growth. Whether a female is impregnated at moulting can therefore affect the frequency of moulting, particularly as several successive batches of eggs can be carried by a

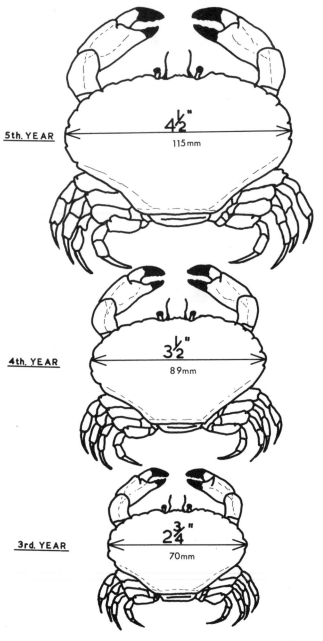

Fig 18 Diagrammatic representation of estimated growth in *Cancer pagurus* based on the results of suture-tagging experiments

Table 17 Details of recaptured crabs which had moulted since being suture-tagged

A. *Increments suggesting one moult.*

Date of recapture	Period before recapture (weeks)	Carapace width (mm)		Increase (mm)	Increase %
		Pre-moult	Post-moult		
		Females			
18.8.59	14	135	155	20	14.8
17.8.60	66	127	145	18	14.2
1.6.60	56	126	154	28	22.2
15.8.60	66	125	146	21	16.8
24.8.59	15	124	149	25	20.2
30.7.59	11	121	151	30	24.8
27.5.60	54	120	146	26	21.7
16.4.60	48	118	146	28	23.7
15.9.60	37	118	140	22	18.6
25.8.60	67	116	137	21	18.1
13.8.60	65	116	133	17	14.6
1.8.60	63	115	140	25	21.7
25.4.60	49	113	135	22	19.5
8.4.60	47	111	136	25	22.5
		Males			
7.7.60	61	129	145	16	12.4
13.7.61	113	119	145	26	21.8
13.7.60	62	118	136	18	15.3
4.4.60	46	117	127	10	8.5
4.4.60	46	116	126	10	8.6
5.4.60	46	114	140	26	22.8
20.6.60	59	112	135	23	20.5

B. *Increments suggesting more than one moult – all females.*

Date of recapture	Period before recapture (weeks)	Carapace width (mm)		Increase (mm)	Increase %
		Pre-moult	Post-moult		
11.7.60	61	127	170	43	33.9
29.7.61	115	120	175	55	45.8
13.5.61	104	116	173	57	49.1

99

female without further moulting and mating. All these factors make it difficult to estimate age accurately in *Cancer pagurus*, and it therefore appears that the relationship between age and size can only be based on average values. Calculations by Williamson (1904) suggested that a crab of 115 mm carapace width (*ie* legal size) '. . . would be not less than three years, nor probably more than four years old.' Pearson (1908) disagreed with this statement and believed that a crab of this size would be nearer five years of age. Meek (1904) estimated that a crab could reach a carapace width of 145 mm (5.75 inches) in the fifth year.

Based on the laboratory and growth experiments made by the present author it is obvious that considerable variation can occur in the frequency of moulting and therefore in the annual growth rate of *Cancer pagurus*. From the available data it appears unlikely that on the east coast of England a crab of either sex would reach 115 mm before the fifth year (*Fig 18*). Edwards and Brown (1967) summarized the growth pattern in the Norfolk fishery, based on average values from laboratory and field studies. According to these results, at the end of its first year of life a crab will have reached a carapace width of 25–30 mm, following three or four moults. During the second year the number of moults may be reduced to two or three and the crab may reach a width of 50 mm. In the third year the crab reaches a size of 70 mm, and a further two years are necessary for it to reach 115 mm. Up to this carapace width growth in both sexes appears similar, but after maturity is reached the annual growth in males becomes reduced, while growth in females is closely related to the reproductive cycle. Due to these factors it is therefore very difficult to accurately age larger members of this species.

According to Bennett (1974), who compared data collected on crab growth from Yorkshire and Norfolk with his results from southwest England, growth in terms of moult increment seems to be very similar in both areas, accepting that there can be significant differences between years and for release areas within or between the northeast or southwest crab fisheries. There are, however, considerable differences in the moult frequency of *Cancer pagurus* from the east and northeast coasts and the southwest. In the southwest male crabs moult more frequently than females, but as stated above the opposite is the case for the east coast of England. In addition, the moult frequency of males on the east and northeast coasts is considerably lower than that of males in the southwest. Conversely, females moult more frequently in the east and northeast than in the southwest, although the observed data for Norfolk and Yorkshire 1959–63, except for one point, are not far removed from the calculated relationship for southwest females. The distinctive difference in the male moult frequency, and consequently annual growth, between the east and northeast and the southwest coasts may be the explanation for the presence of larger male crabs in the southwest of England – a distinctive feature of the population structure in that area.

There are a number of possible explanations for the apparent differences in the growth rates, particularly of males, in the two areas. It may be a genetic difference between the two stocks, or an environmentally controlled or induced difference. No comment can be made on the possibility of some genetic variability. The warmer climate of the southwest of England could result in faster growth. As feeding activity is temperature-dependent the activity of crabs in the southwest is likely to be greater than on the east coast. The food supply may also be better in the warmer climate. The warmer water temperatures in the southwest may be the reason for the prolonged moulting period. This allows crabs to moult twice a year, in the spring and autumn, thus increasing moult frequency. However, one would expect these factors to affect both sexes, whereas it is the males only in the southwest which have a faster growth rate.

Although in recent years the Devon fishery has become the most productive crab fishery in the United Kingdom, the fisheries off the east and northeast coasts of England have been heavily exploited for a longer period. Such heavy exploitation over a long period may be expected to influence population structure and perhaps stimulate density-dependent changes in growth or recruitment rates. However, the difference in growth pattern suggested here is sexually selective – males mainly – and there is a lower growth rate on the east and northeast coasts.

6 Migration

Review of early work

Although fishermen have long known that the European edible crab can move considerable distances (Buckland, Walpole *et al.*, 1877), the first tagging experiments aimed at obtaining definite information on the movements of crabs were made by Williamson (1900) off Dunbar in Scotland. In these experiments the crabs were marked by attaching a numbered brass label to the claw or to a small hole bored in the edge of the exo-skeleton. Crab-tagging experiments using similar tags were also carried out along the east coast of England by Meek (1903, 1905, 1907, 1913, 1914), Tosh (1906), Donnison (1912) and Wright (1931, unpublished). All these experiments suggested that crabs were capable of moving considerable distances. Meek (1913) working in Northumberland pointed out that, in addition to a seasonal inshore and offshore movement by both sexes, female crabs moved in a northerly direction along the east coast of England, whereas males did not exhibit any tendency to migrate north. In all these early experiments tags attached to the claw or exo-skeleton were used, and the results were limited because of the loss of tags following moulting.

Tosh (1906) released claw-tagged crabs in the Yorkshire area but reported that less than five per cent were recovered within ten miles of their original position of release. However, two females travelled in a northerly direction – one from Flamborough Head to Beadnell (Northumberland), a distance of 108 miles in 114 days, and a second which was recovered in Scotland. Meek (1913), working along the Northumberland coast, gave records of ten females which had moved distances in a northerly direction ranging from 25 miles to 160 miles – all into Scottish waters.

In Donnison's (1912) experiment on the Norfolk coast in September 1910, 19 per cent of the tagged crabs were recaptured in the 1911 fishing season (May–September), within a mile of their original position of release. Only one crab, a female which travelled 98 miles north to Flamborough Head in 12 months, moved a long distance.

Williamson also supervised a series of tagging experiments between

1916 and 1924 off the Scottish coast (*Fish. Board of Scotl. Scient. Inv., IV 1929*). These experiments, carried out off Aberdeen and in Berwickshire, involved the release of 2,819 crabs of which 201 (7.1%) had been recaptured prior to 1927. The longest distance between points of liberation and recapture was about 90 miles (Aberdeen to Lossiemouth), this crab having been at liberty for 300 days. Williamson (1940), summarizing these experiments, reported that eight females released off Berwickshire had moved distances ranging from 40–120 miles, in a northerly direction, during a period extending up to four years after release.

Wright (1931) marked and liberated 500 crabs during July and August 1930 within a radius of one mile from Sheringham, Norfolk. Of these, 135 (or 27 per cent) were recaptured within one year of release. The analysis of 125 returns for which the position of recovery was known showed that 46 per cent were recaptured at or near the position of release. There are no records of any females moving northwards out of this fishery, although one did move 22 miles northeast to Cromer Knoll (Wright 1931, unpublished).

More recently the author, in association with Mr M N Mistakidis, released 360 claw-tagged crabs off Norfolk in 1957. Recoveries from these experiments (Mistakidis 1960) totalled 140 (39 per cent) and showed that there was a definite movement towards the land and in a southeasterly direction. No crabs from this experiment were, however, taken outside the Norfolk crab fishery area.

Although all the above experiments demonstrated that *Cancer pagurus,* particularly females, sometimes made extensive migrations, the information available was rather limited, mainly because of the lack of a successful tagging method. The recent development of the suture tag for use in growth studies has also yielded more information on long-term migrations than previous experiments using claw tags.

Local migrations off the east coast of England

From 1959 to 1966 the author carried out a series of tagging experiments along the east coast of England to study the effects of fishing and growth (Edwards 1964, 1965b, 1966a, 1967). Recaptures from these experiments have added considerably to the available information on the migrations of *Cancer pagurus.* This section of the book presents the results from these experiments.

Yorkshire claw-tagging experiments, 1962 and 1964

The intensity of fishing may be estimated from the results of tagging experiments if the tagged animals are distributed randomly among the untagged population (Ricker 1958). Experiments to assess the intensity of fishing on English crab stocks have been undertaken by the author in Norfolk (Hancock 1965; Edwards 1966a) and off Yorkshire (Edwards 1962, 1967). In these experiments claw tags were used because of the low mortality rate associated with their use, but as mentioned previously the tags would be lost at moulting. Even so, some information was obtained on local movements.

During the period 24 March–3 April 1962 a total of 800 claw-tagged crabs were released at 44 stations between Staithes and the River Humber, along the Yorkshire coast. Only crabs of commercial size (carapace width 115 mm) were tagged and each one was marked with a yellow plastic serially numbered disc, attached to the claw by stainless steel wire (*Plate 30*). The recovery of the tags was encouraged by offering a reward of two shillings (10p) for each one returned; the finder was also asked to provide accurate information on the date and position of recapture. By the end of the 1962 season (November) 159 tagged crabs had been returned by fishermen, together with the dates and positions of recapture. The distances moved by the recaptured crabs in the nine months following release

Plate 30   Plastic numbered tag attached by stainless steel wire to a crab's claw

are shown in *Table 18*. These data represent the shortest distance between the points of release and the positions of recapture, and it is therefore possible that the actual distances travelled may have been greater.

Table 18. Returns during the 1962 fishing season of claw-tagged crabs released in March/April 1962 off Yorkshire, grouped by distance between points of release and recapture

Distance (miles)

Sex	1–2	3–4	5–6	7–8	9–10	11–12	13–14	15–20	21–30	Over 30	Total number of crabs returned
Female	31	19	17	7	7	3	3	4	3	2	96
Male	25	23	6	2	1	1	1	3	1	0	63
Total	56	42	23	9	8	4	4	7	4	2	159

Over 60 per cent of the crabs returned were recaptured within five miles of the points of release, and only 13 per cent had moved more than 10 miles during the nine months after release (*ie* April–December 1962).

The recapture positions of tagged crabs which had moved 10 miles or more are shown in *Fig 19*. These movements do not show any greater trend to the north than to the south. Female crabs moved further than males, and two had moved distances of just over 30 miles (both in a northerly direction) in four and six months. The greatest distance travelled by a male was 26 miles in a southerly direction in four months. Three other males also moved 15 miles south in a similar period of time. There is some suggestion that in the autumn some crabs moved offshore; one female released inshore off Filey was later recaptured, in December 1962, by a trawler fishing 25 miles northeast of this port (*Fig 19*). However, the usual concentration of crab-fishing boats within three miles from the shore makes it virtually impossible to collect data on offshore movements.

In this type of experiment the period for which tagged crabs remain free will obviously influence movements. *Table 19* shows the monthly recapture rates for tagged crabs released in March 1962, 80 per cent of which were taken by the end of July, *ie* four months after release.

The substantial decline in the recoveries from August onwards can be associated with the moulting, during which the tags are lost.

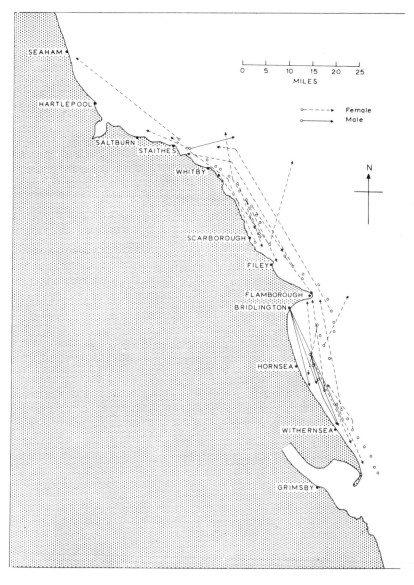

Fig 19   Release stations and movements of 10 miles and over made by claw-tagged crabs released off Yorkshire in March 1962

Recaptures in 1963 totalled only 10 marked crabs, and these results illustrate the limitation of this tagging technique for determining migrations over an extended period.

Table 19 Monthly recaptures of claw-tagged crabs, Yorkshire 1962

(800 crabs released 24 March–3 April)

	Recaptures (numbers)	Percentage
April	6	3
May	38	20
June	62	32
July	55	28
August	21	11
September	3	2
October	5	3
November	1	0.5
December	2	1
Total	193	24

Pearson (1908) and Meek (1931) both reported an inshore movement of crabs in the spring and an offshore movement in the autumn. These migrations were believed to be associated with the moulting and reproductive cycles (Fig 20) but the evidence presented by both authors was sparse. In 1964 the author, by concentrating the release of tagged crabs in one area, collected information on their inshore movements which confirmed that a large-scale immigration of crabs occurred in the spring off Yorkshire.

In April 1964, prior to the commencement of fishing, claw-tagged crabs were released in a section of the Yorkshire coast between Flamborough Head and the Humber (Fig 21). In this experiment 120 pots worked by a hired commercial fishing boat were fished in strings of 30 at distances of $\frac{1}{2}$, $1\frac{1}{2}$, $2\frac{1}{2}$ and $3\frac{1}{2}$ miles from the shore. Fishing in four sections along the coast, the pots were hauled and rebaited on four consecutive days and all legal-sized crabs caught were tagged on the claw with a numbered disc and liberated randomly in the area of capture. During the period 28 April to 2 May, 938 tagged crabs were released at varying depths and distances from the shore. Table 20 shows the number of crabs caught and released at the various distances from shore. Even at this time of year, the highest catches were

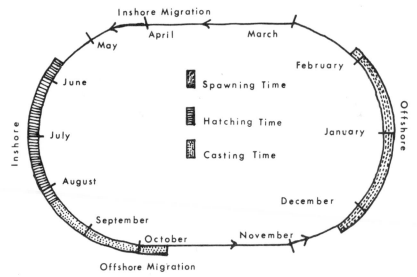

Fig 20   Diagram illustrating the annual migratory cycle of mature crabs. The three processes – casting, spawning and hatching — probably do not take place in one cycle (after Pearson, 1908)

made within half a mile from the shore and, although catches decreased further out to sea, 95 crabs were taken and tagged 3½ miles from shore, which is well out of the normal fishing area.

Table 20 Details of release and recapture of crabs tagged in the spring – 1964

Distance from shore (miles)	Depth (fm)	Numbers caught and released			Numbers recaptured in 1964			
		Total	Females	Males	Total	%	Females	Males
½	3– 4	456	301	155	109	24	61	48
1½	5– 6	229	126	103	66	29	28	38
2½	7– 8	158	70	88	43	27	21	22
3½	10–12	95	41	54	23	24	12	11
Total		938	538	400	241	26	122	119

During the fishing season a constant check was kept on the distribution of fishing and although the fishery was mainly restricted to the area between ½ and 1½ miles from shore, about 25 per cent of the crabs released at 2½ and 3½ miles offshore were later recaptured (*Table 20*).

108

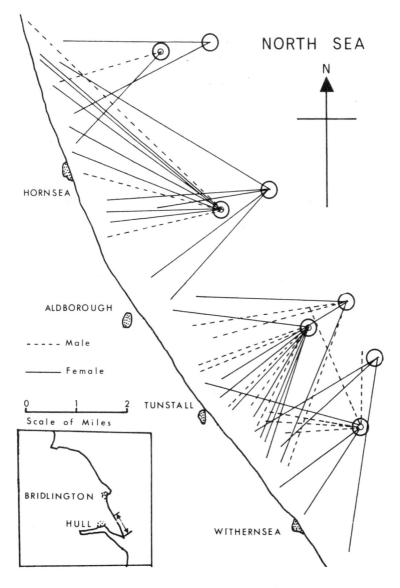

Fig 21 Local movements of claw-tagged crabs released $2\frac{1}{2}$ and $3\frac{1}{2}$ miles off the Yorkshire coast in April 1964 and recaptured during the following four months

Plotted movements during the four months following release (*Fig 21*) also indicate a considerable inshore migration from these offshore stations, although it must be remembered that any offshore movements would not have been traced because the majority of the boats were fishing close inshore. By the end of June, 80 per cent of the released crabs had been recaptured, and during the season which ended in September, 241 crabs were recaptured, representing a recapture rate of 26 per cent. The separate recapture rates for the crabs released at the various depths were similar and ranged from 24 to 29 per cent. However, as shown in *Fig 21*, most of these recaptures were caught inshore in the one fishing area and the majority of the crabs were recaptured within five miles from their original position of release. No records of movements of 20 miles or over were made from this experiment, and the rapid decline in the number of recoveries after July demonstrates again the effects of moulting.

Long distance migrations

Norfolk and Yorkshire suture tagging experiments, 1959–1965
In 1959 and 1960 a total of 1,228 suture-tagged crabs were released off the Norfolk coast (see previous section on growth), 445 (36 per cent) of which have since been recaptured (Edwards, 1965b, 1966a). In these experiments 70 per cent of the crabs later recaptured had moved less than five miles. However, 12 female crabs released in this area were later recaptured by crab fishermen working along the Yorkshire coast and another by a North Sea trawler fishing 60 miles northeast of the Humber. Details of release and recapture are shown in *Table 21* and their positions are plotted in *Fig 22*. It must, however, be remembered that the lines drawn show the shortest possible distance between the points of release and recapture. No males moved more than 10 miles in these experiments. All the females which moved had carapace widths of 140 mm and over when recaptured (*Table 21*), suggesting that only mature females undertook the northward migration.

Recaptures from the 1962 and 1963 suture tagging experiments off Yorkshire, when 2,000 marked crabs were released off this coast, totalled 243 crabs (12 per cent). Of the 136 male crabs recaptured, 95 per cent had moved less than five miles from their original point of

110

Table 21 Details of suture-tagged crabs released off the Yorkshire coast in May 1959 and 1960, which were later recaptured by Yorkshire fishermen

Number	Carapace width (mm)		Position of recapture	Distance moved (miles)	Period before recapture	
	On release	On recapture			Years	Months
1	119	175	3 miles ENE of Scarborough	115	2	2
2	132	No details	1¼ miles SE of Flamborough Head	98	3	2
3	127	170	1 mile off Flamborough Head	100	1	2
4	124	No details	1 mile off Flamborough Head	100	4	0
5	122	No details	6 miles N of Whitby	140	3	0
6	127	No details	½ mile off Mappleton	85	3	2
7	115	173	Filey Bay	110	2	0
8	117	140	Flamborough Head	100	1	4
9	142	168	¼ mile off Hornsea	90	5	0
10	140	165	2 miles ESE of Flamborough Head	98	1	2
11	142	142	½ mile off Hornsea	90	—	11
12	119	175	60 miles NE of the Humber*	95	1	9
13	117	170	½ mile off Flamborough Head	100	4	1

* Recaptured by a Grimsby trawler

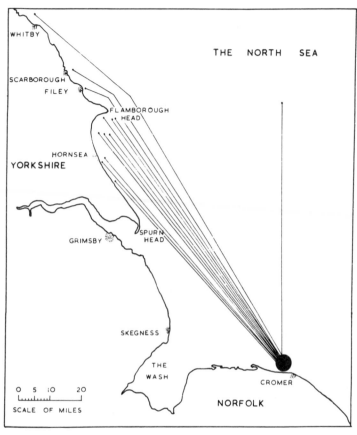

Fig 22 Northerly movement of 13 suture-tagged female crabs released off Norfolk 1959 and 1960

release and only four had moved more than 20 miles. In comparison, 25 of the 197 females recaptured had moved distances of 20 miles and over.

The migrations of 20 miles and over are shown in *Figs 23* and *24*. Once again the movements have been plotted to represent the shortest distance between the positions of release and recapture. Almost all the movements of more than 20 miles were made by females, mostly in a northerly direction. All these crabs had carapace widths over 140 mm (*Table 22*), although several were below legal size when tagged (Edwards, 1967). The longest movement recorded was for a female released off Whitby in 1962 which was recaptured off Fraserburgh, a

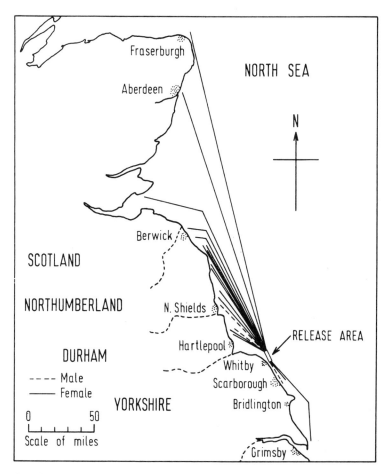

Fig 23 Migrations, 20 miles and over of suture-tagged crabs released off Whitby in 1962 and 1963

distance of 230 miles, after 52 weeks of freedom (*Table 22, Fig 23*). Another interesting recapture was made by a Dutch trawler fishing 40 miles northwest of Ijmuiden in September 1965, which caught a female crab that had been released off Flamborough Head in 1963. When released this crab had a carapace width of 165 mm (6½ inches), and when recaptured (170 miles from its original position of release, 2½ years later) it was found not to have moulted. One female released off Hornsea, Yorkshire in 1962 was recaptured off Sheringham in the following year. This is the first record of a crab moving from

113

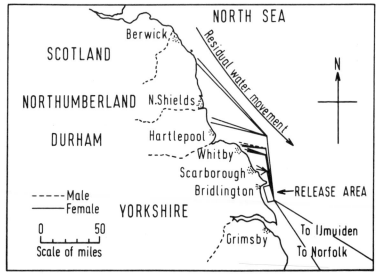

Fig 24 Migrations, 20 miles and over of suture-tagged crabs released off Bridlington in 1962 and 1963

Yorkshire to Norfolk, a distance of about 90 miles, and this is also one of the longest movements in a southerly direction (*Table 22*).

Recaptures from the 1965 growth studies, when 3,400 suture-tagged crabs were released in the Norfolk, Yorkshire and Northumberland crab fisheries (*Fig 25*) have further added to the information on migrations, and crabs were again recaptured at considerable distances from their original position of release. Movements of 20 miles and over from these experiments are shown in *Figs 26–30* and are set out in *Tables 22–25,* together with the periods for which the crabs had been at liberty.

In the Yorkshire experiments a total of 135 females were recaptured and, of these, 39 per cent were later found to have moved distances of 20 miles and over. Of the 53 females which had undertaken such movements only one had moved in a southerly direction, the remainder moving north into north Yorkshire, Durham, Northumberland or Scotland. The longest distance moved was by a female released about 10 miles south of Flamborough Head, which was recaptured off Berwick (Northumberland) 16 months later, having moved 163 miles. For male crabs, out of a total of 144 recaptured off Yorkshire only eight (5 per cent) had moved distances of 20 miles or over and six of these were in a northerly direction (*Tables 23* and *26*).

114

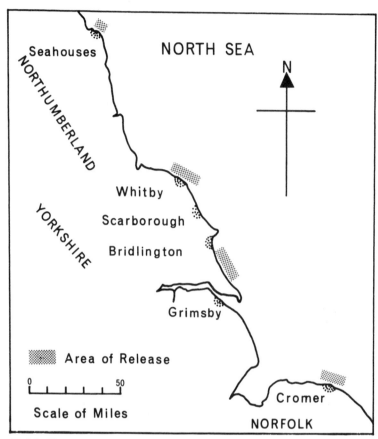

Fig 25 The east coast of England showing areas where suture-tagged crabs were released in May and June 1965

The migration of female crabs out of the Norfolk fishery was not as extensive as that from Yorkshire. Out of a total of 318 females recaptured from the 1965 suture tagging experiments only seven had moved distances of 20 miles or over (*Fig 26*). These females were all later recaptured off the Yorkshire coast. A single male was later found in a trawler's catch close inshore off Sizewell, Suffolk, 57 miles to the south.

In the 1965 experiment over 90 per cent of the Norfolk tagged crabs had moved distances of less than 5 miles (*Table 24*). Results from the 1966 Norfolk releases showed a similar picture and by the end of the 1967 fishing season (*ie* 16 months later) although 274

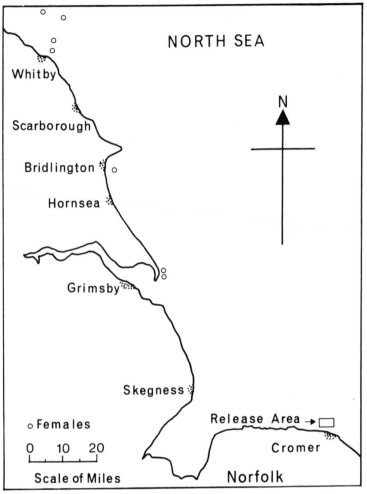

Fig 26   Migrations of recaptured suture-tagged crabs released off Norfolk in June
1965 which had moved 20 miles and over

tagged crabs had been recovered only one, a female, was found to
have moved 20 miles or over.

Only 31 tagged crabs were recaptured from Northumberland but
five of the 10 females recaptured had travelled over 20 miles, all in a
northerly direction into Scotland (*Fig 29*). However, 95 per cent of the
recaptured males and 30 per cent of the females released in this area
had moved less than five miles before being recaptured (*Table 25*).

116

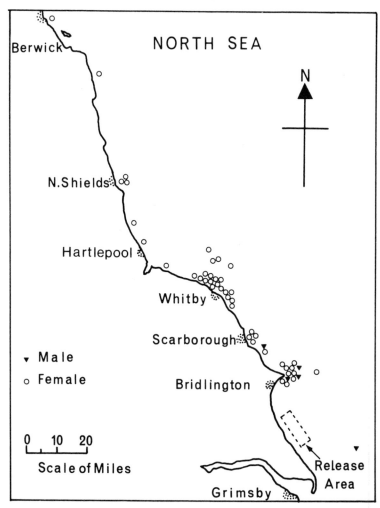

Fig 27 Migrations of recaptured suture-tagged crabs released in the south Yorkshire area in May 1965 which had moved 20 miles and over

The results from the suture tagging experiments made by the author along the east coast of England confirm that there is a definite northerly migration of female crabs along the east coast of England. Meek (1905, 1906) suggested that this migration was connected with the breeding-ecdysis cycle, females which had first moulted migrating along the coast.

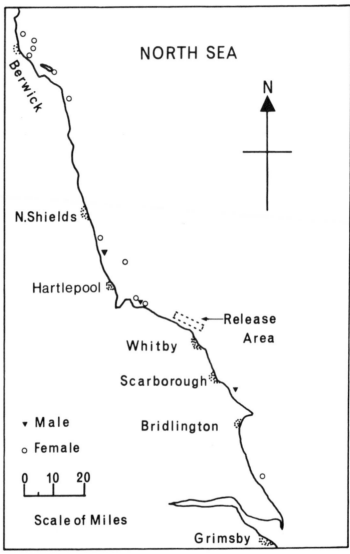

Fig 28 Migrations of recaptured suture-tagged crabs released off north Yorkshire in May 1965 which had moved 20 miles and over

The same view was held by Mason (1965) who reported a northerly movement of tagged females along the Scottish coast. In *Tables 22* to *26* data are presented on the original carapace width and the size on

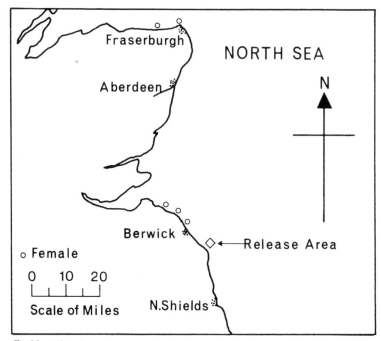

Fig 29 Migrations of recaptured suture-tagged crabs released off Northumberland in
June 1965 which had moved 20 miles and over

recapture from the 1962, 1963 and 1965 suture experiments, although
on some occasions only the tag was returned so the size could not be
determined. All the crabs which had moved 20 miles or over had
carapace widths of at least 130 mm either before or after a moult, and
crabs of this size are known to be mature (Edwards, 1966b). Furthermore, of the 90 females which had moved 20 miles or more, 66 were
known to have moulted. It should also be recorded that, although
some 50 per cent of all the suture-tagged crabs released in these
experiments had carapace widths below the legal size of 115 mm,
there are no records of any of these smaller crabs making long
migrations.

These results also showed that tagged crabs recaptured during the
year in which they were released had moved only short distances, and
the majority of the long distance movements were by female crabs
recaptured in the year following release, although a few crabs were
taken in the second and third years after release. According to Meek
(1913) mature females tend to move out of the inshore fishing areas to

119

offshore spawning areas in the autumn. As described earlier in this account, 'ripe' females had been observed by the author six to seven miles off Whitby in depths of 25–30 fathoms. Spawning occurs during the winter, and in the following spring and summer the hatching females are common in the catches taken inshore. This northerly migration of female crabs along the east coast of England and Scotland is therefore believed to be part of an offshore movement in the autumn followed by a return to the shallower inshore waters in the spring, when the eggs are hatched. On the western side of the North Sea, near the English coast, the residual current is in a southerly direction, so that crabs moving in a northerly direction are walking against the residual current. When the larvae from these crabs are released in the late spring or early summer they are free-floating and drift southward for about 30 days (Pearson 1908) before they finally settle to the bottom and assume the adult shape. This northerly migration of adult female crabs which occurs along the east coast of England is therefore believed to be part of an inherent behaviour pattern which results in the southerly transport of the larvae to an area suitable for their survival. The results also confirm Meek's observation that males do not make extensive migrations but move at random. The results from these recent suture tagging experiments suggest that this northerly migration of female crabs is greater than was shown by earlier workers.

The author (Edwards 1967), presenting carapace width measurements of the Yorkshire crab stocks, showed that the average size range of crabs landed in this area increases from the southern part of the fishery (Grimsby–Bridlington) northwards (Whitby). When males and females are considered separately it is seen that the size range of male crabs caught varies little from south to north. In contrast, females caught in the southern part of the fishery are smaller than those caught further north (*Fig 30, Table 27*). The occurrence of a greater proportion of large female crabs in the catches landed at Whitby and Scarborough is believed to be associated with the northerly migration of crabs along the Yorkshire coast.

Studies in southwest England

Dr David Bennett and Mr Clive Brown also collected interesting results on migration from the Channel area (Bennett and Brown 1976). They tagged some 9,000 crabs by the suture method during the years 1968–71. In their experiments recaptures from both inshore and

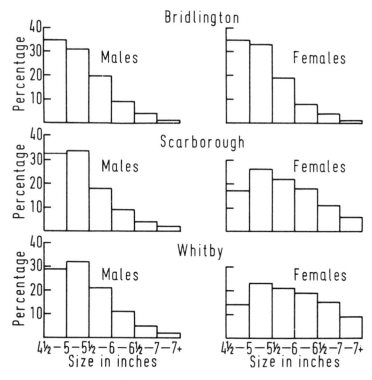

Fig 30 Proportions of the Yorkshire crab catch in the various carapace width size groups – based on data collected 1961–64 (after Edwards 1967)

offshore tagging experiments have shown that some crabs, particularly females, make extensive movements, mainly west or southwest down the English Channel (*Fig 31*). Of those recaptured, between 10 and 32 per cent moved 10 miles or more, and the majority of these (58 to 80 per cent) moved westwards or southwestwards down the Channel. Although a few males made movements over 10 miles, the majority of movements were made by female crabs. The maximum distance covered was 155 miles. The concentration of female crabs with developing or well developed ovaries in the autumn off south Devon may well be the result of a breeding migration. There is probably a large, but possibly diffuse, stock consisting mainly of females over a large area of the English Channel, and extensive migratory movements, mainly down the Channel, produce complex stock relationships. The movement of male crabs is limited and stocks of

121

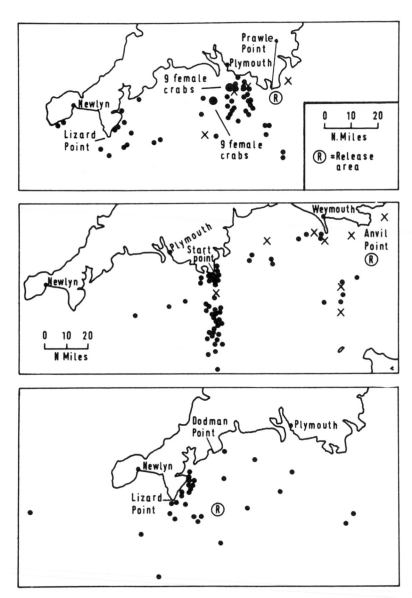

Fig 31 Recapture positions of male (×) and female (●) crabs which moved 10 or more miles from the release areas in southwest England. (A) Prawle Point, October 1971; (B) 12 miles south of Anvil Point, July 1973; (C) 20 miles south of Dodman Point, July 1973. R = release position (after Bennett and Brown, 1976)

Table 22 Distances of 20 miles and over travelled by suture-tagged crabs released during growth experiments off the Yorkshire coast in 1962 and 1963

Nly = Northerly   Sly = Southerly   N R = No record

Size on release (mm)	Sex	Period before recapture (weeks)	Minimum distance moved (miles)	Direction travelled	Moulted
141	Female	55	88	Nly	No
105	Female	46	47	Nly	Yes
133	Female	52	230	Nly	NR
146	Female	62	82	Nly	No
135	Female	48	86	Nly	Yes
142	Female	48	126	Nly	Yes
115	Female	55	85	Nly	NR
183	Female	11	28	Nly	No
114	Female	62	93	Nly	Yes
159	Female	16	25	Nly	No
185	Female	53	21	Sly	No
178	Female	16	51	Nly	No
112	Female	40	46	Sly	Yes
116	Female	48	20	Nly	Yes
128	Female	40	57	Nly	NR
136	Female	47	46	Nly	Yes
110	Female	44	83	Nly	Yes
118	Female	41	24	Nly	Yes
112	Female	41	31	Nly	Yes
137	Female	47	40	Nly	Yes
122	Female	47	43	Nly	Yes
128	Female	41	34	Nly	Yes
129	Female	51	40	Nly	Yes
142	Female	130	170	Ijmuiden	No
120	Female	55	90	Sly	Yes
164	Male	61	40	Nly	No
91	Male	60	21	Sly	Yes
140	Male	40	22	Nly	No
120	Male	52	25	Nly	Yes

Table 23 Distances of 20 miles and over travelled by suture-tagged crabs released off Whitby, N. Yorkshire in 1965

Sex	Size on release (mm)	Size on recapture (mm)	Period before recapture (months)	Minimum distance travelled (miles)	Direction of movement
Female	157	tag only	17	110	N
Female	115	145	16	88	N
Female	114	tag only	No details	46	N
Female	108	137	15	49	S
Female	142	172	13	110	N
Female	131	tag only	14	100	N
Female	117	145	16	113	N
Female	118	148	14	20	N
Female	118	147	11	23	N
Female	124	148	24	100	N
Female	109	136	11	20	N
Male	135	154	16	25	N
Male	159	159	5	41	N
Male	104	135	17	32	S

Table 24 Distances of 20 miles and over travelled by suture-tagged crabs released off Norfolk 1965 and 1966

Sex	Size on release (mm)	Size on recapture (mm)	Period before recapture (months)	Minimum distance travelled (miles)	Direction of movement
Female	114	176	19	140	N
Female	144	173	14	68	N
Female	111	170	40	140	N
Female	117	150	14	68	N
Female	116	tag only	25	110	N
Female	124	150	28	145	N
Female	112	166	28	144	N
Male	112	tag only	5	57	S

124

Table 25 Distances of 20 miles and over travelled by suture-tagged crabs released off the Farne Islands, Northumberland 1965

Sex	Size on release (mm)	Size on recapture (mm)	Period before recapture (months)	Minimum distance travelled (miles)	Direction of movement
Female	146	177	12	30	N
Female	198	198	29	160	N
Female	131	158	28	188	N
Female	148	tag only	16	37	N
Female	155	155	5	20	N

Table 26 Distances of 20 miles and over travelled by suture-tagged crabs released off Tunstall, S. Yorkshire in May, 1965

Sex	Size on release (mm)	Size on recapture (mm)	Period before recapture (months)	Minimum distance travelled (miles)	Direction of movement
Female	125	155	15	78	N
Female	119	147	16	25	N
Female	145	145	15	52	N
Female	117	149	11	58	N
Female	138	169	14	96	N
Female	114	169	31	58	N
Female	117	146	13	26	N
Female	104	159	27	65	N
Female	107	134	14	41	N
Female	127	159	13	61	N
Female	130	157	18	72	N
Female	98	tag only	29	38	N
Female	124	156	12	62	N
Female	126	159	22	40	N
Female	104	161	19	66	N
Female	108	135	14	41	N
Female	125	154	17	63	N
Female	133	162	10	86	N
Female	134	160	13	25	N
Female	115	139	13	25	N
Female	131	156	14	106	N

Sex	Size on release (mm)	Size on recapture (mm)	Period before recapture (months)	Minimum distance travelled (miles)	Direction of movement
Female	120	tag only	16	25	N
Female	134	tag only	17	163	N
Female	146	176	13	26	N
Female	121	151	5	26	N
Female	120	186	30	106	N
Female	134	tag only	24	25	N
Female	129	157	12	70	N
Female	135	170	39	68	N
Female	125	155	12	35	N
Female	123	150	13	60	N
Female	125	154	8	60	N
Female	139	166	14	65	N
Female	145	175	30	60	N
Female	131	163	14	106	N
Female	138	168	14	24	N
Female	119	145	16	59	N
Female	119	144	14	53	N
Female	121	151	13	40	N
Female	115	171	24	58	N
Female	114	140	14	25	N
Female	110	163	43	60	N
Male	110	140	12	26	N
Male	126	160	28	25	N
Male	107	135	13	25	N
Male	116	145	37	35	N
Male	99	132	9	32	S

cock crabs are therefore probably localised and dependent upon recruitment from local sources.

The results from tagging experiments in Scottish waters, along the northeast and east coasts, and in southwest England demonstrate that edible crabs can undertake considerable migrations, often in excess of 100 miles. These migrations are associated with the offshore movement of females for spawning. Male crabs rarely moved far from the point of release.

Table 27 The percentage of the crab catch landed between certain size limits for three Yorkshire ports – 1961–64 (After Edwards 1967)

Port	Sex	Size-group (inches)						Percentage 6 inches and over
		$4\frac{1}{2}$–5	5–$5\frac{1}{2}$	$5\frac{1}{2}$–6	6–$6\frac{1}{2}$	$6\frac{1}{2}$–7	Over 7	
Bridlington	Male	35	31	20	9	4	1	14
	Female	35	33	19	8	4	1	13
Scarborough	Male	33	34	18	9	4	2	15
	Female	17	26	22	18	11	6	35
Whitby	Male	29	32	21	11	5	2	18
	Female	14	23	21	19	15	8	42

7 Present state of the stocks

In recent years there have been some variations in landings in the three main crab fishing areas of England, *ie* the coasts of Northumberland, Yorkshire, the southwest coast of Devon and Cornwall and the small, but productive fishery off Norfolk. For example, crab landings along the northeast coast have declined: in 1962 some 40 per cent of the English crab catch was landed at ports in Northumberland and Yorkshire, but by 1975 only 20 per cent of the English crab catch came from this northeast coast. In contrast, landings in the southwest, especially in the area between Plymouth and Brixham, have steadily increased, and by 1975 some 70 per cent of the crab catch was landed there (*Fig 32*). The Norfolk fishery is a fairly stable one with annual landings representing on average about 10 per cent of total annual English catch.

Landings on the northeast coast have not declined because of a shortage of crabs but because of a change in the local pattern of

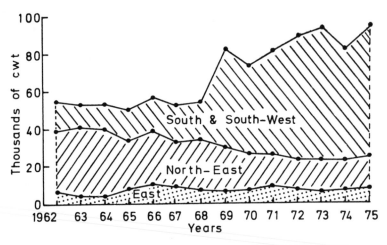

Fig 32 Crab landings for 1962–1975 from the main fisheries. Northeast = Northumberland to Lincolnshire; East = Norfolk to Essex; South and Southwest = Kent to Land's End, Cornwall

128

fishing. This has come about because many northeast coast fishermen have changed from crab fishing to other more remunerative fisheries such as lobstering, trawling or salmon fishing. This has led to a decline in the number of pots fished for crabs and has resulted in lower landings. Even so, the stocks of crabs off this coast are still fairly heavily exploited at certain times of the year and no large increase in long-term catches is likely, even with further increases in effort.

The crab fishery in the southwest, particularly off Devon, has flourished in recent years with increased demand and landings. Stocks on the inshore grounds (*ie* within 12 miles of the coast) have been heavily exploited, and in order to maintain high catch rates, Devon boats have had to work further offshore. Recent studies by Ministry scientists (Bennett and Brown, 1976) suggest that the Devon crab stocks are not capable of withstanding any further large increases in exploitation, and recommendations have been put forward that the minimum landing size for both males and females should be increased in this area. This proposed management action does however require fundamental changes in the national legislation in order to allow regional minimum sizes (at present all sizes apply to the whole of the UK) and different minimum sizes for male and female crabs.

Some areas of the country do offer scope for further exploitation: there are ample stocks of crabs off parts of the Welsh coast and they remain virtually unexploited. Similarly in the eastern English Channel the large stocks could withstand some fishing.

Crabs are plentiful on all coasts of Scotland but they are mainly fished off the east coast. However, Shetland and, to a lesser extent, Orkney now account for about 30 per cent of the total landings. On the east coast the traditional fisheries along the coasts of Grampian, Tayside, Fife, Lothian and Borders are all generally heavily exploited and no large increases in landings can be expected. On the west coast of Scotland ample stocks of crabs exist but they are only lightly fished at the moment.

8 Future of the fishery

As described in this lecture British fishermen have caught crabs around our coasts for centuries. What is their future in this fishery? For years British fisheries scientists have been stressing the fact that many stocks of crabs around this country remain under-exploited and are capable of further exploitation. Even so, British landings of crabs show little signs of any increase. All this points to a distinct lack of encouragement for fishermen to land crabs – in other words a poor market, or in some areas no market at all.

Why doesn't the crab compete with other shellfish such as the lobster and *Nephrops* ('scampi')? Crab meat is a favourite of the housewife – imported crab meat in the canned form constitutes an important fishery import, valued as high as £1.5 million in some years. While it is white meat that is imported, some brown body meat also enters the country and is used by our paste manufacturers. If we import this meat why can't we produce it ourselves?

Processing centres

Unless there is a crab processing plant in his area, most fishermen do not have a market outlet for large quantities of crabs. These plants are expensive and, furthermore, crab fishing usually coincides with the holiday period when better paid and more pleasant work is available at our coastal resorts. Because of these staff problems some shellfish processors have taken fewer and fewer crabs and have concentrated on other types of fish or shellfish. This poor demand by processors is often reflected in the low prices paid to fishermen to land crabs, and they consequently increase their efforts to catch lobsters or change to trawling or other types of fishing.

The Torry Research Station, Aberdeen has been considering methods of extracting crab meat mechanically. This would reduce labour costs and could make meat extraction more profitable. Trials are now going ahead in this country with an American machine which separates white meat from the claws, legs and bodies of crabs. Originally designed for species such as Dungeness crab, there appears no reason why it cannot be used to process the European crab which has a good yield of excellent flavoured meat. The US system incorporates a decanting centrifuge in which the meat and shell are separated in brine of a particular specific gravity after initial comminution. Initial results have shown that yields can be increased considerably over the normal hand operation.

The process first involves the butchering of the cooked whole crab. The claws and legs and parts of the body (with the brown meat removed) are then fed into a chopping machine. The chopped meat and shell is then fed into the centrifuge in a stream of brine. This brine plus the meat and shell is then separated by centrifugal force – the shell sinks in the brine and the meat floats and can be collected easily.

The widespread use of machinery of this type in the UK could help to increase landings and utilise stocks not at present exploited.

How can the demand for crabs be increased? People living on or near the coast appreciate a nice crab salad. Further inland, crabs on the fishmongers slabs are items of curiosity rather than a source of culinary delight.

A good case can be made for up-grading the image of the crab in Britain. This could be done with the aid of women's magazines and TV cookery programmes. Few young housewives know how to clean and prepare a crab. Some sort of promotion scheme could increase the demand.

At present most crab meat is eaten with salads, or as pastes or spreads. The utilization of crab meat, especially the brown body meat, into 'crab cakes', 'crab sausages' or other attractive fish products could further help to stimulate demand. The question of large-scale canning could also be considered. Torry Research Station is currently looking at this as well as other possible methods of utilising crab meat.

Exports

If the market outlet in the UK does not improve, the opportunity to export should not be missed. We now have closer ties with our EEC neighbours, several of which have a considerable appetite for shellfish. The recent development of an export market in live spider crabs (*Maia squinado*) is an example of what can be done with some effort. Here a demand from Spain and France has resulted in a regular supply of live spider crabs being sent over to Europe – kept alive in well-boats or in 'vivier' lorries. Prices paid for these spider crabs, once considered a pest by fishermen in southwest England, now exceed those paid for the common edible crab, although some are now being sent alive to Spain thus increasing the overall price in Devon.

Improved marketing

We know the resources are available, fishermen can catch and handle the species, and technological methods are being developed to extract the meat mechanically. If crab fishing is to improve economically, then better marketing will have to be used to obtain a realistic price for what is a prime product.

This aim can only be achieved by moving away from the 'live' crab sales with its problems of limited storage life, seasonal limitations and the problems of transporting such a perishable product. Freezing could be the answer. Deep-frozen cooked crab is a good product, but professional marketing would be required to persuade both an export and home market to accept frozen crab when they are used to fresh crab.

If the meat separation process described earlier becomes a practical proposition and is introduced into the UK, increased crab meat production could force processors to find new outlets. This in itself could expand marketing opportunities. The future for our crab fishermen looks better now than it has done for some years.

Bibliography

Bell, T., 1853. *A history of the stalk-eyed crustacea*. John Van Voorst, London.

Bennett, D. B., 1970. Crab tagging in south-west England: an interim report. *ICES, C.M. 1970, Doc. No. 13* (mimeo).

Bennett, D. B. and Brown, C. G., 1970. Crab investigations in south-west England. *Shellfish Inf. Leafl., Fish. Lab. Burnham-on-Crouch, No. 18* (mimeo).

Bennett, D. B., 1974. Growth of the edible crab (*Cancer pagurus* L.) off south-west England. *J. Mar. biol. Ass. U.K* 54 (4) p 803–823.

Bennett, D. B. and Brown, C. G., 1976. The crab fishery of south-west England. *Lab. Leafl., Fish. Lab. Lowestoft*, No. 33, 11 pp.

Broekhuysen, G. J., 1937. Some notes on sex recognition in *Carcinides maenas* (L.). *Archs. néerl. Zool.*, Vol. 3: pp 156–164.

Brown, C. G. 1975. Norfolk crab investigations. *MAFF Lab Lt No 30*.

Buckland, F., 1875. *Report on the fisheries of Norfolk especially crabs, lobsters, herring and the Broads*.

Buckland, F. and Walpole, S., *et al.*, 1877. *Reports on the Crab and Lobster Fisheries of England and Wales, of Scotland and of Ireland*. H.M.S.O., London.

Burkenroad, M. D., 1947. Reproductive activities of Decapod Crustacea. *Am. Nat.*, Vol. 81, pp 392–8.

Carlisle, D. B. and Dohrn, P. F. R., 1953. Studies on *Lysmata seticaudata* Risso (Crustacea Decapoda). 2. Experimental evidence of a growth – and moult accelerating factor obtainable with eyestalks. *Pubbl. Staz. Zool. Napoli*, Vol. 24: pp 69–83.

Carlisle, D. B. and Knowles, F. G. W., 1959. Endocrine control in Crustaceans. *Camb. Univ. Monogr. exper. Biol.*, Vol. 10: pp 120. (Cambridge Univ. Press).

Cheung, T. S., 1966. The development of egg-membranes and egg attachment in the shore crab, *Carcinus maenas* and some related Decapods. *J. mar. biol. Ass. U.K.*, Vol. 46, pp 373–400.

Cunningham, J. T., 1898. On the early post-larval stages of the common crab (*Cancer pagurus*), and on the affinity of that species with *Atelecyclus heterodon*. *Proc. Zool. Soc.*, Pt. II, p 204.

Donnison, J., 1912. *Report on crab investigations*. Eastern Sea Fisheries Committee, Boston.

Drach, P., 1939. Mue et cycle d'intermue chez les crustacés decapodes. *Ann. Inst. Océanogr. Paris*, Vol. 9, pp 103–391.

Edwards, E., 1962. Yorkshire crab investigations. *Lab. leafl. Fish. Lab. Burnham-on-Crouch* (New Series), No. 3, 12 pp.

Edwards, E., 1964. The use of the suture-tag for the determination of growth increments and migrations of the edible crab (*Cancer pagurus*). *ICES, C.M. 1964*, Doc. No. 42 (mimeo).

Edwards, E., 1965a. The crab fishery of England and Wales. *Wld Fisg*, London, Vol. 14, No. 3, pp 57–61.

Edwards, E., 1965b. Observations on the growth of the edible crab (*Cancer pagurus*). *Rapp. P.-v. Reun. Cons. per. int. Explor. Mer*, Vol. 156: pp 62–70.

Edwards, E., 1966a. The Norfolk crab fishery. *Lab. leafl. Fish. Lab. Burnham-on-Crouch* (New Series), No. 12: 23 pp.

Edwards, E., 1966b. Mating behaviour in the European edible crab (*Cancer pagurus* L.). *Crustaceana*, Vol. 10, Pt. I, pp 23–30.

Edwards, E., 1966c. Further observations on the growth of the edible crab (*Cancer pagurus*). *ICES, C.M. 1966*, Doc. No. 17 (mimeo).

Edwards, E., 1967. The Yorkshire crab stocks. *Lab. leafl. Fish. Lab. Burnham-on-Crouch* (New Series), No. 17, 34 pp.

Edwards, E., 1971. A contribution to the bionomics of the edible crab (*Cancer pagurus* L.) in English and Irish waters. PhD Thesis, National University of Ireland, 123 pp.

Edwards, E. and Early, J. C., 1967. Catching, handling and processing the edible crab. *Torry Advisory Note*, No. 26, 17 pp.

Edwards, E. and Brown, C. G., 1967. Growth of crabs in the Norfolk fishery. *Shellfish Inf. leafl., Fish. Lab. Burnham-on-Crouch*, No. 8 (mimeo).

Edwards, E. and Meaney, R. A., 1968. Observations on the edible crab in Irish waters – pt. I. *Resource Rec. Paper, Irish Sea Fish. Bd*, Dublin (mimeo).

Gundersen, K. R., 1963. Tagging experiments on *Cancer pagurus* in Norwegian waters. *Annls. biol., Copenh.* 18, pp 206–208.

Gurney, R., 1942. *Larvae of decapod Crustacea*. Roy. Society, London, 306 pp.

Hallbäck, H., 1969. Swedish crab investigations. Some preliminary results. *ICES, C.M. 1969*, Doc. No. 27 (mimeo).

Hancock, D. A., 1965. Yield assessment in the Norfolk fishery for crabs (*Cancer pagurus*). *Rapp. P.-v. Réun. Cons. perm. int. Explor. Mer*, Vol. 156, No. 13. pp 81–93.

Hancock, D. A. and Edwards, E., 1967. Estimation of annual growth in the edible crab (*Cancer pagurus* L.). *J. Cons. perm. int. Explor. Mer*, Vol. 31, pp 246–264.

Hartnoll, R. G., 1969. Mating in the Brachyura. *Crustaceana*, Vol. 16, pp 161–181.

Holdsworth, E. W. H., 1874. *Deep-sea fishing and fishing boats*. Edward Stanford, London.

Lebour, M. V., 1928. Larval stages of the Plymouth Brachyura. *Proc. zool. Soc. Lond.*, pp 473–560.

Mason, J., 1965. The Scottish crab tagging experiments, 1960–61. *Rapp. P.-v. Reun. Cons. perm. int. Explor. Mer*, Vol. 156, No. 12.

Meek, A., 1903. The migrations of crabs. *Rep. Northumb. Sea Fish. Comm.* 1903, p 33.

Meek, A., 1904. *The crab and lobster fisheries of Northumberland*, ibid, 1904.

Meek, A., 1905. *The crab and lobster fisheries of Northumberland*. I The value of protection II The migration of crabs. ibid., 1905, p. 26.

Meek, A., 1907. *Migrations of crabs* ibid., 1906, p. 26.

Meek, A., 1913. *The migrations of crabs*, ibid., 11, p. 13.

Meek, A., 1914. *Migrations of the crab*. ibid., 111, p. 73.

Mistakidis, M. N., 1959. Preliminary data on the increase in size on moulting of the edible crab, *Cancer pagurus*. *ICES, C.M. 1959*, Doc. No. 52 (mimeo), 2 pp.

Mistakidis, M. N., 1960. Movements of the edible crab (*Cancer pagurus*) in English waters. *ICES, C.M. 1960*, Doc. No. 88 (mimeo).

134

Nordgaard, O., 1912. Faunistiske og biologiske i akttageleser ved den biologiske station i Bergen. Trondhjem. *Kge. Vid. Selsk, Skr.,* No. 6 1911, pp. 1–58.

O'Céidigh, P., 1962. The marine Decapods of the counties Galway and Clare. *Proc. Royal Irish Acad.,* Vo. 62, pp. 151–175.

Pearson, J., 1908. Cancer (the edible crab). *Mem. Lpool. Mar. biol. Comm.,* No. 16, 263 pp.

Ricker, W. E., 1958. Handbook of computations for biological statistics of fish populations. *Bull. Fish. Res. Bd Can.,* No. 119, 300 pp.

Thomas, H. J., 1958a. Lobster and crab fisheries in Scotland. *Mar. Res.,* 1958, No. 8, 107 pp.

Thorson, G., 1950. Reproductive and larval ecology of marine bottom invertebrates. *Biolog. Rev.,* Vol. 25, pp. 1–46.

Tosh, J. R., 1906. Crab migration experiments. *Interim Rep. North-Eastern Sea Fish. Committee.*

Van Engel, W. A., 1958. The blue crab and its fishery in Chesapeake Bay. Part I – reproduction, early development, growth and migration. *U.S. Fish. Wildlife Serv.,* Vol. 20, 17 pp.

Waterman, T. H., (Editor), 1960. *The physiology of Crustacea. Volume 1. Metabolism and growth.* The Academic Press, New York and London, 670 pp.

Williamson, D. I., 1956. The plankton of the Irish Sea 1951 and 1952. *Bull. Mar. Ecol.,* Vol. 4., No. 31, pp. 87–144.

Williamson, H. C., 1900. Contributions to the life-history of the edible crab (*Cancer pagurus,* Linn.). *Rep. Fish. Bd Scotland,* Vol. 18 (3) pp. 77–143.

Williamson, H. C., 1904. Contributions to the life-histories of the edible crab (*Cancer pagurus*) and of other Decapod Crustacea. *Rep. Fish Bd Scotland,* Vol. 22, (3) pp. 100–140.

Williamson, H. C., 1940. *The crab fishery.* pp. 1–64. Aberdeen Journals Ltd.

Wright, F. S., 1931. *Report on the results of crab marking experiments at Sheringham and district in 1930.* Ministry of Agriculture and Fisheries, London (mimeo).

Yonge, C. M., 1955. Egg attachment in *Crangon vulgaris* and other Caridae. *Proc. Roy. Soc. Edinburgh,* Vol. 65, pp. 369–400.

Index

Index

Other books published by Fishing News Books Limited
Farnham, Surrey, England

Free catalogue available on request

A living from lobsters
Advances in aquaculture
Aquaculture practices in Taiwan
Better angling with simple science
British freshwater fishes
Coastal aquaculture in the Indo-Pacific region
Commercial fishing methods
Control of fish quality
Culture of bivalve molluscs
Eel capture, culture, processing and marketing
Eel culture
European inland water fish: a multilingual catalogue
FAO catalogue of fishing gear designs
FAO catalogue of small scale fishing gear
FAO investigates ferro-cement fishing craft
Farming the edge of the sea
Fish and shellfish farming in coastal waters
Fish catching methods of the world
Fish farming international No 2
Fish inspection and quality control
Fisheries oceanography
Fisheries of Australia
Fishery products
Fishing boats and their equipment
Fishing boats of the world 1
Fishing boats of the world 2
Fishing boats of the world 3
Fishing ports and markets
Fishing with electricity
Fishing with light
Freezing and irradiation of fish
Handbook of trout and salmon diseases
Handy medical guide for seafarers
How to make and set nets
Inshore fishing: its skills, risks, rewards

141

International regulation of marine fisheries: a study of regional fisheries
    organizations
Marine pollution and sea life
Mechanization of small fishing craft
Mending of fishing nets
Modern deep sea trawling gear
Modern fishing gear of the world 1
Modern fishing gear of the world 2
Modern fishing gear of the world 3
Modern inshore fishing gear
More Scottish fishing craft and their work
Multilingual dictionary of fish and fish products
Navigation primer for fishermen
Netting materials for fishing gear
Pair trawling and pair seining – the technology of two boat fishing
Pelagic and semi-pelagic trawling gear
Planning of aquaculture development – an introductory guide
Power transmission and automation for ships and submersibles
Refrigeration on fishing vessels
Salmon and trout farming in Norway
Salmon fisheries of Scotland
Seafood fishing for amateur and professional
Ships' gear 66
Sonar in fisheries: a forward look
Stability and trim of fishing vessels
Testing the freshness of frozen fish
Textbook of fish culture; breeding and cultivation of fish
The fertile sea
The fish resources of the ocean
The fishing cadet's handbook
The lemon sole
The marketing of shellfish
The seine net: its origin, evolution and use
The stern trawler
The stocks of whales
Training fishermen at sea
Trawlermen's handbook
Tuna; distribution and migration
Underwater observation using sonar